# AMERICAN POISON

## ALSO BY DANIEL STONE

*Sinkable*

*The Food Explorer*

# AMERICAN POISON

*A Deadly Invention
and the Woman Who Battled
for Environmental Justice*

# DANIEL STONE

DUTTON

**DUTTON**

An imprint of Penguin Random House LLC
1745 Broadway, New York, NY 10019
penguinrandomhouse.com

DUTTON and the D colophon are registered trademarks of
Penguin Random House LLC.

Permissions appear on page 339 and constitute an extension
of the copyright page.

Book design by Shannon Nicole Plunkett

LIBRARY OF CONGRESS CATALOGING-IN-PUBLICATION DATA
has been applied for.

ISBN 9780593473627 (hardcover)
ISBN 9780593473641 (ebook)

Printed in the United States of America

1 3 5 7 9 10 8 6 4 2

The authorized representative in the EU for product safety and compliance
is Penguin Random House Ireland, Morrison Chambers, 32 Nassau Street,
Dublin D02 YH68, Ireland, https://eu-contact.penguin.ie.

*For Micah and Jonah*

# CONTENTS

It would be tragic if, many decades from now,
it were recognized from accumulated evidence that
large segments of populations in ours and other nations
had suffered needless disability and torment because
early warning signs . . . went unheeded.

—Clair Patterson, geochemist, 1966

# Author's Note

This is a story about poison. Not the kind that kills you quickly, like cyanide or arsenic. But the kind that kills you slowly, that embeds itself in your blood, lungs, and brain and quietly erodes your ability to function until eventually, after a series of odd pains, strange coughs, and mysterious side effects, your body gives up.

Poison of this sort isn't usually called poison—at least, not at first. It's called a new miraculous substance or a wonder cure. Something that can make food taste better, or houses stay warmer, or pipes resist corrosion. The things that seem like a good idea at the time, even divine, but when they're examined more closely, over decades, they prove not only faulty but obvious disasters from the start.

Luckily, it's not all bad news. This is also a story about Alice Hamilton. Don't despair if you've never heard of her. Most of our storytelling around early environmentalism centers around a few well-worn names like Alexander von Humboldt, Teddy Roosevelt, and Rachel Carson. One day, while preparing a lecture for my environmental science students at Johns Hopkins University, I saw a footnote—a literal footnote—about a woman in the 1920s who defined a new era of environmentalism, then called public health and safety. Functionally, it was the same idea as today's global battles: Humans were doing dangerous things, often at the expense of other humans, and in the process making neighborhoods, cities, and the world dirtier and sicker.

Hamilton's introduction to environmentalism did not come from a holy experience in nature, like John Muir's in Yosemite or Thoreau's at Walden. Instead, she spent her formative years in Chicago, where she began noticing curious things happening in nearby factories and warehouses. Men were coming home from work sick. They were working with materials like lead, rayon, and tungsten. Most seemed to lack basic protections like gloves or uniforms. And almost all of them were immigrants with families to support.

The American excitement of this era, embodied in breakthrough inventions like the Wright brothers' flying biplane and the mass-produced automobile, didn't spread to the millions of Americans who shoveled coal into the boilers that kept America running. The handymen, the factory workers, and the uncountable laborers. As long as the economy kept growing, there was little to be gained by asking questions about things as dull as safety and health that might slow things down.

Alice Hamilton noticed this imbalance. And, notably, she might have taken the more common path and simply looked away. She was a white female doctor raised in a wealthy family that owned an entire city block in Fort Wayne, Indiana. But the marginalization she faced as a woman in her field made her sympathetic to those who were marginalized even more. She decided to work at Hull House, a Chicago settlement house created by Jane Addams that offered basic services to immigrants and the poor.

Other people, however, had no such sympathy for the workers making American progress happen. If health mattered, scale and profit mattered more. A growing number of capitalists and industrialists simply wanted to innovate solutions and get rich in the process. This was the spirit driving progress in oil, steelmaking, railroads, and food production. But nowhere was it more present than in the making of cars, which promised to change everything about American life.

Thomas Midgley embodied this American spirit, a sense of possibility and a willingness to pursue a goal at any cost. I won't spoil your introduction to Midge except to say that he was a driven and brilliant inventor who decided early in life that he wanted to change the world. Undoubtedly, he succeeded. Sometimes the cost of innovation is small, and sometimes, as you'll see in this story, it's big.

The cost is also personal. I'm sheepish to admit that, even though I teach environmental science, it took becoming a dad before I considered how personal pollution can be. Part of that is privilege: I never had to grow up next to railroad tracks or on a busy street. Working on this book made me sensitive to the micro-poisons all around us: in our carpeting, our water, the plastic toys our kids put in their mouths. Even though governments make sweeping guarantees for things like clean water and safe medicines, the reality can be flimsier. Scratch below the surface, and it becomes clear that the mechanisms to protect us from things that will make us sick have blind spots, especially when driven by the incentives of capitalism.

This is not a sad story. It's a shocking one, I'll grant, but ultimately it's a story of two incredible Americans and their fight to build conflicting worlds. It's also a story about how change starts and progress happens—how a transformative invention came into the world, and how a pioneering woman started a movement to fight it for a century to follow. Where that leaves us now—well, let's catch up at the end and we can all take a guess.

**Daniel Stone**
Santa Barbara, California

# Part One
# THE SIDE DOOR

# 1

*Alice Hamilton in 1915*

## Cambridge, Massachusetts — 1919

For as long as America had existed, one of the country's top centers of knowledge and science circled a large patch of grass two miles west of Boston. America's first university had begun here in 1636 as a school named for a Puritan clergyman who donated half of his life savings and four hundred books to start a college in the new colony in the Massachusetts Bay.

In the three centuries since its founding, Harvard University had expanded in size and space and grown John Harvard's initial endowment to almost $10 million. But the patch of grass remained, now with walking paths and black iron benches, all surrounded by large buildings where the finest professors gave the finest lectures to the finest students, all of them men.

Alice Hamilton sat on a bench on the morning of March 15, 1919, amid snow flurries, looking at the burned crater of Dane Hall, a building in the center of campus that had been destroyed in a spectacular fire the year prior. She had just come from a meeting with the dean of Harvard's medical school, David Edsall, who, outside of his dean capacity, happened to be an expert in the effects of extreme heat in enclosed spaces.

Hamilton's meeting with Edsall was the final step in a long-running conversation that took place by typewritten letters that shuttled between Cambridge and Chicago, where Hamilton lived. The discussion in these letters was timid and noncommittal, at least at first, because it was about something no dean at Harvard had ever thought about, much less acted upon. Edsall wanted to recruit Hamilton to join the faculty of Harvard. If Edsall was successful, Hamilton would be the first woman ever considered for the faculty ranks.

Hamilton sat quietly watching male students walk in and out of the buildings. She was short, five foot four, with light brown hair parted down the middle. She wore a beige turtleneck with a brown blazer, an outfit she confided to a friend that she favored in professional settings because it looked sharp but bland, and not overly feminine. The snow was melting and signs of life had returned to the university quad. In front of her, two students tossed a ball, their breaths forming clouds of vapor. Other students sat on the steps of Sever Hall with their heads buried in books.

Hamilton's path to that bench had been uneven, even rocky at times. She was born in 1869 into a wealthy family, one of the richest in Fort Wayne, Indiana, and had grown up on an estate covering three city blocks that was owned by her grandfather, who had made a fortune in land speculation and railroad promotion on Indian lands. By the time Hamilton was four, her father, Montgomery Hamilton, pushed Alice and her three sisters to direct their own education by reading books. Montgomery believed that all education should be rooted in learning languages—Latin, French, and German—which would open a wider array of books for his girls to learn everything else themselves.

When Hamilton turned seventeen, she went away to boarding school at Miss Porter's School in Farmington, Connecticut. Miss Porter's fed her hunger for more formal learning that Hamilton needed. The curriculum was created by Sarah Porter in 1843 to teach young girls to be good Christians first and, second, to be good wives and mothers "whose influence would shine like a saving light in a domestic world." Porter was morally opposed to excess, corruption, and selfishness and worked to make her school an epicenter of reform to move girls forward on the educational and social track. She had no interest in frivolous activities or distractions. "There may be dancing in the backroom of the wing," the school's handbook read in the 1850s. "But Miss Porter wishes that there should be no waltzing."

For Hamilton, the straitlaced education worked. She graduated with a well-formed goal of becoming a doctor who could help people, which launched her almost immediately into medical school in Michigan. A year later, at twenty-four, she had a medical degree and, not long after, a job studying pathology at Northwestern University in Chicago. To be young and unattached was a great pleasure for Hamilton. Her days were full of research and correspondence, and her evenings were entirely hers to roam Chicago and attend lectures, plays, and musical performances—a cosmopolitan life in the Gilded Age for a woman who had no interest in bothersome distractions like men, children, or making money. As the decades passed, she developed an expertise and, with it, a reputation.

The original spark that brought Hamilton to Harvard was a conversation between the owners of Boston's three biggest department stores: Filene's, Gilchrist's, and Jordan Marsh. Two years earlier, in 1917, one of the store owners began to notice that the turnover of his workers for health-related reasons was so high that it was beginning to be a drag on his business. He mentioned this to the other owners, who admitted to having the same problem. The cause of so much sudden illness was a costly mystery.

Together, they agreed a survey should be done about the general health of department store workers and the reasons for their illnesses—not necessarily for the sake of the workers but for the owners to have more reliable employees and thus keep up their receipts. They proposed the idea to the Massachusetts Retail Trade Board, and a year later the board came up with $50,000 for a five-year study of retail workers.

Harvard was the obvious venue for such a study. Its medical school was the most authoritative in the country, and it happened to be an hour from downtown Boston by foot and twenty minutes by automobile. Edsall was delighted when a member of the Retail Trade Board offered him the money to do the study. All he needed was somebody to design and run the investigation. A search was done of professors at other universities who could be brought to Harvard. But the field of industrial medicine, as it was clumsily called, was less than a decade old, with few seasoned experts.

Edsall remembered an article he had read years earlier in the *Boston Globe* about how Dr. Alice Hamilton, a Chicago physician, conducted a similar occupational investigation in 1911 commissioned by the governor of Illinois that looked at the conditions of workers in lead factories in Chicago and found detailed and actionable ways to improve worker health. In Edsall's mind, this made Hamilton the leading candidate for the job, and also the only one.

When he tried to get in touch with her, Edsall was surprised to learn that Hamilton had a richer résumé than just the Illinois survey. In her thirties, Alice Hamilton had single-handedly created and become the de facto head of an entirely new field of medicine rooted in exactly this kind of research. The dangerous trades, she called them—the factories and smelters that manufactured materials, fuels, and consumer products, and the workers who operated these facilities.

In a time of booming production and thousands of buzzing factories across America, Hamilton was one of the first people to wonder about the health of the workers. And when she did, she found countless instances of men who toiled with scarcely any protections at all. No uniforms, no gloves, and—for those who worked with especially gruesome chemicals like mercury and lead—no respirator masks. "Dangerous" was a gentle word for these jobs. Often they were deadly.

Hamilton was sympathetic toward the men who held these jobs. They were usually European immigrants, almost all of them living on pennies. Most spoke broken English at best, which left them marginalized. Holding no political or economic power, they had little choice but to accept any work they could find, no matter how paltry or perilous.

What was more, Hamilton lived in Chicago, a city newly infamous for exactly this kind of worker peril. Most worker drudgery was kept hidden from view to protect rich people from confronting such ghastly tales. But that changed in 1906 when Chicago's slaughter yards and packinghouses made international news. A socialist writer named Upton Sinclair had embedded himself for two months in Chicago's Packingtown, as it was called, where he observed unspeakable squalor, including workers living twelve to a room in fetid boarding-houses and packing floors so dirty that even a scrape of one's knuckles could cause an infection that could kill a man in days. Sinclair's work was eventually published as a novel he titled *The Jungle* that, despite containing composite characters and some fictional details, included stomach-churning scenes that shocked the Victorian sensibilities of the upper class. In one of the worst, Sinclair wrote of men who accidentally fell into hog vats and were boiled alive, then were fished out, ground up, and sent out to the world as Durham's Pure Leaf Lard.

In the early 1900s, the law protected people making money, not the workers with hand sores, broken bones, and breathing difficulties. Even children were swept into the gears of American industry far more than in European countries that had long-standing child labor laws. In 1900, 2 million U.S. children still worked in mills, mines, fields, and factories. Two attempts by Congress to protect children from exploitation were overturned by the Supreme Court until, in 1938, a law called the Fair Labor Standards Act finally stuck. Immigrants, however, had no such heartstring appeal. The lines of desperate workers that formed outside factories were all but proof to business owners that nothing needed to change.

The worst workplace dangers were usually the hardest ones to see. Any man knew he could accidentally cut his finger off while using a sharp knife to debone a cow. Or he could eventually end up with a burn if he shoveled bricks into a furnace. But it was harder to link the effects of breathing mercury vapors to future lung disease, or the lack of coverall uniforms to crippling body pains that would land a man in bed and eventually the grave.

Hamilton's talent was finding the connections. Her usual

process was to interview sick men in their homes and at their bedsides—or, worse, to interview their widows about their husbands' last days—and then compare the details of their afflictions to the symptoms of other men. With enough data, she could find patterns. "Shoe-leather epidemiology," someone called it.

She loved this work. According to one biographer, she enjoyed the one-to-one relationship between cause and effect so rare in medicine. Her work was both scientific and personal, and it fit squarely with her childhood goal that she could somehow, in some way, "make the world better."

The work also suited her. The one factor that made her especially good was knowing how to use her gender to her benefit. A man showing up for a surprise inspection at a factory would get turned away for fear that he was an agent for the government or a competing company. But a woman posed no threat at all, especially a five-foot-four soft-spoken one who was not angry or rude but relentlessly polite and curious. What harm could she bring? Being underestimated had its benefits.

In the weeks before their meeting at Harvard, Edsall appealed to Hamilton's sympathy for the toiling American worker. "I am writing . . . to ask you whether you would consider taking charge of this work," Edsall wrote to Hamilton two days after Christmas in 1918. "It is apparent, I think, that there is a very large opportunity for public service in a field that has been very inadequately studied."

He also toyed with Hamilton's sense of prestige, making an opportunity no woman had ever been given. "It is the first time that the proposition has ever come up to have a woman appointed to any position professorial or other in the University." He said he thought it would be a "large step forward in the proper attitude toward women," particularly at a time when most colleges didn't admit women even as students, let alone faculty. Schools that did subjected them to such crippling discrimination that, less than fifty years prior, several philanthropists endowed a series of women-only colleges in the Northeast where they could study without harassment.

Flattery could boost her ego, but Hamilton was not in need of a job. In fact, what she needed was less work. For the past five years Hamilton had lived in feverish motion between Chicago, Baltimore, and Philadelphia. Several months before, she had agreed to oversee a survey for the Department of Labor on stonecutters who were coming down with high rates of tuberculosis due to a new mechanical cutting device that cut stone ten times faster but also produced ten times as much dust. A woman who had even basic competence in medical matters surprised many men, and when they learned she was untethered to a husband or children, professional opportunities often followed—sometimes too many of them.

Plus she already had money, enough to suit her, and all of it hers. She bought a winter home with her sisters beside the Connecticut River near New Haven and would travel to Michigan in the summer. She accepted an invitation in 1915 to the International Congress of Women in The Hague in the Netherlands and was often asked to do work for the National Research Council and the Public Health Service. The rest of the time, she simply went wherever anyone invited her to research, study, or speak.

When the letters started appearing from Harvard, Hamilton at first ignored them. She thought it was a hoax or a condescending appeal for a woman to make the men feel virtuous. When they kept coming, she wrote to her sister Edith, to ask how she should respond. Many other women might have jumped at the chance to join America's most exclusive boys' club. But Hamilton and her sister were not ordinary women.

Besides, Hamilton didn't *want* to do the department store study, which sounded boring. She had plenty of work and her schedule was full. And she didn't want to move to Boston. It was too far, it was too cold, and she had no friends there.

Finally, in January 1919, she responded to Edsall, saying no thanks. She could not commit to a full-time study, and she was not the right person to conduct hundreds of interviews with department store workers, who were mostly women inclined to small talk and chitchat. She was accustomed to interviewing men, ones who worked in factories, mines, and smelters. And besides, a survey of department store workers—not exactly the most marginalized of people—would be funded by the industry, which all but guaranteed corporate handholding. She had been through this before. An industry only wants to uncover problems it can easily solve, not ones that make it look bad.

How about half-time? Edsall responded. And forget the department store work. Hamilton could come to Harvard simply to teach. She could instruct a class in industrial poisonings, he said, or, frankly, whatever she wanted to do. September to February. Free housing. And $2,000 a year, a sum $200 more than the average Harvard salary.

Hamilton could not believe how successful her efforts at negotiation had been. And that Edsall had folded so completely.

"Isn't this wonderful?" she wrote in a laughing letter to Edith. "He did away with every single objection and made it as easy as possible to accept." The idea had started to grow on her. As long as she wasn't tied solely to Harvard, her affiliation with the school would validate her other work and the general acceptance of women in science. It would also let her perform more profound and probing work investigating grittier places like factories and mines.

Her feelings about moving, however, were unchanged. "What am I going to do for six months each year in Boston?" she wrote Edith. "It appalls me to think of it."

Hamilton had a hunch that a woman's reception at Harvard would not be as warm as Edsall vowed it would be. Edsall was enthusiastic about her arrival, as was the university's president, A. Lawrence Lowell, after Edsall explained to him the competitive advantage Hamilton would bring to the medical school as the leader of a nascent field. Lowell and Edsall were also growing self-conscious that Harvard's refusal to admit women as students or faculty was beginning to make the school stand out for the wrong reasons, at least compared to Cornell, which had been accepting women since the 1870s.

But the group of men who made up the Harvard Corporation, the effective board of the university, were squarely opposed to diluting the bastion of free thought with a woman who might distract from their serious work. During the closed-door meeting of the council that discussed Hamilton's appointment, one professor swore loudly that he believed Hamilton joining their ranks would be the end of Harvard.

After the council begrudgingly voted her through, her new colleagues demonstrated their disdain with gritted smiles. On her first day, a professor congratulated her and then asked her to confirm that she would not take advantage of her right to enter the male-only Harvard Club. She obliged. He also asked if she would request football tickets. She assured him she would not. Almost weekly she had to sidestep similar indignities to protect her male colleagues' fragile egos. In the spring of 1920, after she had taught her first semester, she received an invitation to the commencement ceremonies with the awkward disclaimer printed at the bottom that "under no circumstances may a woman sit on the faculty platform."

The snub didn't bother her. It "seemed a bit tactless," she later wrote, "but I was sure it was not intentional."

Her reward for walking softly was feeling that she had vaulted to a new height and proved that other women could too. "I'm not the first woman who ought to have been called to Harvard," she told a reporter for the *Boston Globe*. Other papers declared, "The Last Citadel Has Fallen," and that Hamilton had finally arrived to "break down the sex barrier."

Hamilton was indeed a symbol, but she wasn't alone as a woman breaking barriers. Everywhere one looked, there were signs that women were finally making headway in their push to be taken seriously as more than just wives and mothers. It hadn't been long since activists pushed breakthroughs in birth control, women's education, and fair pay. The path to more social change was through government policy, and the only way to dismantle sexist policies was to replace stubborn male lawmakers with more progressive ones, especially if a few of them were women.

Finally, in 1918, the year Harvard thought it might be time to hire a woman, the relentless energy of the political organizers broke through. An election ushered in a new Congress sympathetic to women's right to vote. In the spring of the following year, the Nineteenth Amendment passed and was quickly ratified by thirty-six states, giving 26 million American women a reason to celebrate a right that many couldn't believe they had to fight so hard for.

And here, at Harvard, was a woman who managed to break a barrier not by demand but by invitation. She was good at her work and, without portraying a large sense of pride, she knew it.

Despite the pageantry and the newspaper interviews, the sexism and the polite hostilities, Hamilton was no longer tired. She felt the energy and purpose of a person at the top of her field who had been granted the opportunity to rise even higher. If it ever occurred to her that almost any woman alive would envy her opportunities, she didn't dwell on it for long. She had work to do, and she was excited to do it.

This was convenient, because her work would soon become national news, and eventually her name would circle the world.

The story, however, would not be about her. People didn't want to read about a scientist who researched poisons. They wanted to read about the poison.

# 2

*Thomas Midgley around 1917*

## Dayton, Ohio — 1916

It was snowing the morning Thomas Midgley Jr. walked into work for his first day as a junior scientist in Dayton, Ohio. The snow had been coming in truckloads, dumping more than two inches an hour, a quantity common for Ohio in the winter, but at such speed that the *Dayton Daily News* called it "a greeting from the Eskimos."

Midgley had wanted to be an inventor since he was a boy, mostly because his father was also an inventor. Tom Sr. was an avid cyclist who worked for a series of rubber companies making tires; at one of them he invented the first demountable tire rims for automobiles, which won him the kind of small-town acclaim that got him stopped at the general store. Tom Sr. was the kind of father who gave his son riddles and brain puzzles, urging him to think of questions sideways to find an answer no one else had thought of.

It helped that young Midgley devoured almost every news article and book about the famous inventors of prior decades. The turn of the twentieth century was an exciting time for children in America, but especially boys, who could see men who looked like them accomplishing great things. Thomas Edison in New Jersey, Alexander Graham Bell in Nova Scotia, and the Wrights in North Carolina. And for every major breakthrough like flying machines, telephones, and electric lights, there were hundreds of smaller ones like mousetraps, ice-cream scoops, curtain rods, and flyswatters. The papers carried tales of all of them, along with backstories of the brilliant minds locked in thrilling races of discovery, leading any reader, especially an impressionable young man, to believe that even the smallest effort would result in new discoveries.

This combination of influence and timing all but predetermined that young Midgley would become an inventor himself. And sure enough, his first big chance came when he was fourteen, while playing baseball in high school. Midgley and a teammate noticed that if a pitcher spit on a baseball, he could make it spin faster, and if it spun faster, it would curve sharply as it approached the batter. This was the well-known principle of a curveball. But Midgley wondered if an even stickier substance would create an even sharper curve. He went home and made a list of the stickiest things he could think of, like maple syrup and honey. After some trial and error, he landed on the sap of the slippery elm tree. Eventually, smearing sticky tree goo on baseballs was adopted by players at other high schools who grew up to play professional ball.

Fresh off a success, Midgley's next goal was to *work* as an inventor, the kind who got paid to solve high-stakes puzzles and make appliances, knickknacks, and time-savers of the future.

Midgley believed that the top advances of the future would be in automobiles. It hadn't been long since even *seeing* an automobile was an occurrence rare enough to be brought up at the dinner table. The world's first automobile, known as the Benz Patent-Motorwagen, was built in Germany in 1886, and after several iterations the first motorized wagons found their way to the United States. More than simply providing personal transportation, which was formerly a luxury for the elite, automobiles symbolized progress and the dawn of something new.

This was a welcome change, to say the least. Horse-drawn carriages—the dominant mode of transportation—were onerous and dirty. At the end of the nineteenth century, there were more than 100,000 horses living in New York City, and millions more across America, each depositing up to thirty-five pounds of manure each day. Enormous piles of urine-soaked dung on sidewalks were not the only things left behind; the rotting corpses of abandoned horses also littered the streets.

Horse*less* carriages, as early automobiles were called, provided the quickest way to replace animal power. They promised freedom and adventure, plus the exciting prospect of reshaping society and what it meant to travel. Advertisements urged Americans to "See America First," meaning before anyone else. Within a few years, automobiles clogged the roadways leading into Yosemite Valley and the Great Smoky Mountains, each one followed by a thick cloud of exhaust.

This didn't bother Midgley. Automobiles everywhere, seemingly all of a sudden, was what made them so exciting. He enrolled in Cornell University in 1908 to study mechanical engineering. He spent his days learning about mechanics, thermodynamics, and materials science. Most of the students hated the required chemistry courses, but Midgley loved the concept of atoms and how they bonded together to make every substance imaginable, like baseballs, wood bats, and tree sap. In the evenings and on weekends, he sketched molecules in a small notebook like a diary. One day he decided to start carrying a copy of the periodic table with him everywhere he went, a habit he would continue for decades.

This made Midgley smart but not yet qualified for greatness. He landed a few odd jobs as a junior assistant or paper pusher. But his father was impatient to ignite his son's enthusiasm for automobiles, and he secured Tom Jr. a job at the tire company where he himself was a senior officer. Suddenly, Midgley, at twenty-five, had the title of "chief engineer."

Midgley's inflated title and his father's backing gave him leeway to work on whatever he wanted. He decided to use his time at the tire company to do something entirely unrelated to tires. Instead, he wanted to make a hydrometer that would indicate whether an automobile's cooling system contained enough alcohol to prevent it from freezing in the winter. Midgley knew enough about Edison, Bell, and the Wrights to know that the best solution was usually the simplest idea with the fewest parts. And so he didn't invent an elaborate hydrometer with hinges and gears but merely two balls of different densities. When they were put into a radiator with the right amount of alcohol, one of the balls would float and the other would sink. Such simplicity impressed his father and the company bosses, evidence of his privilege to work off script and be celebrated at the top levels for it. Midgley's success with the hydrometer might have led him to more quirky ideas had the company not run out of money and lost market share to a fast-growing rubber company founded in Akron in 1915 called General Tire.

Luckily, America was awash in other companies making advances in automobiles. And nowhere were there more of these companies than Ohio, the center of the automotive industry until Henry Ford yanked it up to Michigan. Ohio had access to raw materials like coal, iron, and ore and easy proximity to transportation networks by railroad and waterway, with more factories and foundries than anywhere else, plus a workforce skilled in steel and heavy machinery. Ohio was also central, giving the state's companies access to markets in the East and Midwest and as far as California. A boy wanting to work with automobiles was as lucky to be from Ohio as an aspiring banker would be to grow up on Wall Street.

Had he tried, Midgley could have gotten an interview at any company he wanted. He was young, smart, and enthusiastic. It didn't hurt that he had a successful father who could vouch for him. But Midgley wanted to work for one company, and particularly the man who ran it. And if he could, it would be the kind of job that would determine the destiny of his life.

Charles F. Kettering had never heard of Thomas Midgley. But he decided he'd give the young man an interview anyway.

Kettering was a celebrated start-up founder of his day, a man who had early success and rode the wave as far as it would take him. At thirty-three, Kettering founded the Dayton Engineering Laboratories Company (Delco), which became successful almost overnight thanks to Kettering's groundbreaking invention of an electric starter motor. The traditional starter crank was noisy and dirty, the most onerous part of operating an automobile. It was dangerous, too: Sometimes the crank failed to disengage from the shaft and would start turning so hard it would knock the cranking person off their feet—or, worse, break their thumb or shoulder. Kettering's electric starter opened driving to those who lacked the strength—or the Victorian sensibility—to crank a pipe covered in soot. For making this discovery at the precise moment that Americans were falling in love with the automobile, Kettering was celebrated in newspapers and on magazine covers as the century's newest innovator. He also got rich, thanks in large part to Delco's first big customer, another newly formed firm known as the General Motors Company, which in 1911 placed an order for self-starters worth ten million dollars.

Midgley was enchanted by Kettering's success. Kettering demonstrated the exact kind of puzzle solving that drove Midgley to want to be an inventor. But Midgley was more fascinated by Kettering's style of working. In media profiles, Kettering, then thirty-nine, was described as possessing a sense of curiosity and charisma. He had an inventive spirit and a ceaseless determination. Unlike the go-it-alone self-centeredness of someone like Edison, Kettering seemed to value collaboration and teamwork. He wanted good people working for him—people who could think for themselves and tell him when he was wrong.

"It's important work," Kettering said when Midgley came for his interview in early 1916. The automobile industry was more than about just making automobiles, he said. Companies like Delco and General Motors had the ability to change American life from top to bottom, transforming what people were capable of and who would benefit from the advances of the new century. This was an opportunity for acclaim and profit, but it was also a risk: Failure would just as easily lead to ridicule. Kettering didn't want Midgley, then twenty-five, thinking that Delco was a playground for the idle enthusiast. He wanted hard work and results.

Kettering liked that Midgley was young. Just thirteen years separated them, but Kettering saw "Midge," as he called him in that first conversation, more as a son than a peer. Youth helped a man see problems with fresh eyes unencumbered by the mental constraints that came with decades of experience and too much knowledge. And indeed, Midgley later validated this hypothesis with a study of his own. He found a list of history's most outstanding inventions and, scouring their inventors' biographies, he discovered that 90 percent of them had been made by people younger than forty.

Midgley fit right in. He wore a brown suit with slicked brown hair combed to the right. He flashed a warm smile to Kettering that had won him friends since childhood. The moment the two men sat down, they already had a collection of colleagues in common and stories from the inner workings of the nascent industry. They laughed about how Midgley had worked for a short time at the National Cash Register Company in the same department where Kettering many years earlier had also been a young inventor. They chided each other with cajoling and banter the way men of that era did who looked the same, talked the same, and came from similar backgrounds. After an hour, it went without saying that Midgley got the job. Kettering extended a handshake and welcomed Midgley on board.

Five days later, in heavy snow, Midgley walked up to the door of Delco's headquarters at 329 East First Street in Dayton. It was a boxy building, six stories tall, with Kettering's office on the second floor.

Midgley pulled at the door, but it was locked. He tried to pull it harder, then gave it a loud knock. Was he at the right place? Wasn't this where he had come for his interview?

Just then a man popped his head out the window. "Hey,

bud, the architects just stuck that door on to look right to fool people. You get in on the side." The man was Kettering, which made Midgley smile. It was a literal metaphor of Kettering's signature style of inventiveness: If the usual way doesn't work, try a side door.

Dayton was small, much smaller than Columbus, where Midgley had grown up. Dayton was a stereotypical American small town: quaint, friendly, and filled with poor roads. A few years before, the town had expanded its network of electric railcars powered by burning coal, which had turned Dayton from a feeble dot on the map into a bustling metropolis of 130,000 people. Dayton's mayor promised the railcar network would help the city grow and increase the number of stores and eateries on Jefferson Street and Monument Avenue. But even he didn't know how dramatic the growth would be. New neighborhoods sprung up in every direction of the city so fast that nobody had time to think about planning or zoning. Building was so haphazard that streets met at odd angles. This wasn't a problem when streets were solely for streetcars and pedestrians, but it became an issue when automobile drivers had to share the roads too.

Midgley's move to Dayton to start a new job was a big undertaking. His childhood ambition to do exciting work was complicated by the fact that, by 1916, Midgley was a family man with a wife and two small children who would have to move to Dayton as well. His wife, Carrie Reynolds, was from Pleasantville, an even smaller Ohio town than Dayton. Midgley's father was responsible for that connection, too, particularly after Tom Sr. realized that his son seemed to be more interested in drawing molecules than talking to girls. The marriage "took," as was said at the time, and within six years Midgley went from being a bachelor to supporting a family of four, living the all-American dream of having one daughter and one son.

Midgley's expected salary of $1,050 per year would be enough to support them all. But that meant he needed to keep the job, and keeping it required that he do good work. It wasn't lost on Midgley that in a booming era of industrial growth and competition, his boss defined good work as work that made money.

Kettering showed Midgley where he would sit: not in an office but in a laboratory that looked and smelled more like a factory. The floor was a mix of concrete and dirt, and the air was stale and damp. A few years earlier, in 1913, the Miami River had flooded downtown Dayton, and almost every building, including the Delco headquarters, was still rebuilding.

But these were mere aesthetics to Midgley. For a young man who had always wanted to be a bona fide inventor for a leading American company, that first day at Delco had the feeling of a dream.

Walking around the shop floor, he met engineers who were working on every conceivable part of an automobile. One man was testing new spark plugs and ignition coils to increase, as he called it, "fuel efficiency." Another man was connecting electric lights to the starter system to power headlights and taillights and, in a true flourish of innovation, to illuminate inside the passenger cabin. Others were formulating bigger batteries that could meet the growing electrical needs of automobiles.

Within a few days, Kettering gave Midgley his first assignment. After hearing Midgley's tale of the hydrometer at the rubber company, he asked Midgley to build a different sort of hydrometer, one that measured the degree of charge in a storage battery.

Midgley didn't know the first thing about batteries, but he knew he needed to learn fast. He went to the lab and started to tinker.

Three days later, he appeared in Kettering's office and set

down a contraption on the desk. It was the hydrometer Kettering had asked for. But, like his last hydrometer, rather than build an appliance entirely from scratch that would sit beside the battery, he had surmised that all he needed was a membrane of celluloid that would sit inside the battery liquid with a dial to point to a number signifying the battery's charge. And to prevent it from corroding or combusting, he thought to coat it in powdered pumice stone and sodium cyanide, a chemical that could withstand the cold temperatures of an Ohio winter and was also cheap to buy. Midgley demonstrated how it worked. On battery after battery, it measured the charge perfectly.

The following year, Delco filed a patent for Midgley's scrappy invention, which Midgley officially described as a "new and useful improvement." A year later, a refined version of his hydrometer began to appear in all new automobiles.

Midgley made his first test look easy. But he knew there would be harder ones to come. Future challenges would require deep corners of his brain and the far reaches of the periodic table.

For now, he couldn't help but show a flare of excitement that he had completed his first task for Kettering—and that while he was pushing the boundaries of what was possible, he was also improving American life in ways large and small. After he and Kettering inspected every detail of the hydrometer, Midgley looked up at Kettering and said he was ready for more.

"What do you want me to do next, boss?"

# 3

*Hull House when Alice Hamilton arrived in 1897*

## Chicago, Illinois — 1899

Every morning began at exactly six thirty with the paper carrier, who stuffed copies of the *Chicago Times-Herald* and the *Chicago Record* under the front doormat. Sometimes he'd meet a tired mother bringing her baby to the nursery or a milk deliveryman replenishing empty bottles.

By seven o'clock the bakery would begin to fill the hall-ways with the smell of white bread and morning rolls, which would alert the volunteers sleeping upstairs that it was time to head down to the coffeehouse before breakfast disappeared.

In the breakfast hall, hands grabbed the rolls and fruit as fast as they were set out. People flooded down the stairs and in the front door. First came the factory workers whose shifts started early, then the librarian, the doctors, and the schoolchildren. Last would come the tired homebodies who had stayed up late the night before in language class or sewing club.

This was how Hull House came alive every morning: first slowly, and then with the boisterousness of a train station where people with little likelihood of running into each other would find themselves in line for coffee or to use the bathroom, all while they shouted past each other in German, Italian, Russian, and Polish.

A decade earlier, in 1889, a restless young woman had this vision of a loud and messy place where all were welcome. The woman, Jane Addams, had read about a settlement house in London called Toynbee Hall, whose purpose had been to expand the bounds of culture, class, and education by inviting people together whom society tended to separate. After Addams graduated from the new Smith women's college in Massachusetts, she imagined all the ways the world could be improved with more compassion and empathy. Seeing no one else working to change it, she decided to do it herself.

It helped that when Addams was twenty-one, her father, a former railroad executive and Illinois state senator, died and left her $50,000. Fifty thousand dollars was sufficient seed money for Addams's goal. She rented a dilapidated old mansion on Halsted Street that was formerly owned by a real estate developer named Charles J. Hull. The rent was cheap because it was in the heart of a sprawling slum amid factories, run-down saloons, dirty grocery stores, and butcher shops. The wooden shanty huts that housed extended families had no water or sewage pipes. The odors from the garbage overlapped with the smells of cooking grease, human waste, and chickens and pigs in pens.

Addams loved the energy of the neighborhood, the way children would dart through muddy alleys playing chase or hide-and-seek. But she also saw the way constant stress was limiting the people's ability to improve their lots, or at the very least find steady sources of money and food.

The house needed work. Addams spent a year restoring the mansion's carved walnut moldings and marble fireplaces. She wanted it to be beautiful, and a place that would make people feel important—and invited.

In September 1889, she started to fill the house with privileged and educated young people—all women at first—who she believed would support and build relationships with locals and, in so doing, strengthen and integrate the city's fabric. "A bridge between the classes," she called it. It wasn't just that the people needed help improving their wretched conditions or fighting government corruption. The women volunteers could help with that. But even more, Addams believed they needed compassion to find work, skills, and purpose that made them feel valued. Before it had a name, what Addams envisioned was social welfare.

She filled the calendar with classes and clubs to help people foster new skills and interests. Six classes a day, thirty each week, from singing and needlework, to the *Odyssey* and Shakespeare, to Lincoln and the Young Citizens Club, to physics and geometry, and even to additional languages like French, Latin, and German.

The experiment was often messy. Over the next ten years, it became clear that the stress of being poor and powerless could inflame already fraught cultural relations. Almost daily, tiny misunderstandings could land a Czech and a Ukrainian at each other's throats without being able to communicate what they were even fighting about. Racial divides could be even worse. Despite Addams's purported vow that everybody was welcome, the few Black families in the neighborhood who came to Hull House for a class or a club meeting were usually treated coldly. Addams might have welcomed them in, but the white immigrants tended to premise their participation on the exclusion of the Blacks, and if people stopped coming, Addams feared the whole experiment would unravel.

But Addams kept at it. And when she did, she noticed that money poured in from donors and do-gooders who wanted to help without dirtying their hands. People all over Chicago were concerned about filthy and overcrowded tenements. Donating to Hull House offered a way to calm their consciences from the comfort of their homes. Addams received invitations to speak all over town and eventually across the country about her successful work putting young do-gooders to work in service of the poor and marginalized.

It was a compelling argument, especially to one young woman with light brown hair sitting in the audience at one of Addams's speeches in Fort Wayne, Indiana. In the spring of 1897, Alice Hamilton sat in a church pew listening intently. It was as though Hamilton was seeing an older version of herself—a woman who turned her drive to be helpful into a sustaining and satisfying project that was improving people's lives.

"We have in America a fast-growing number of cultivated young people who have no recognized outlet for their active faculties," Addams often repeated in her speeches. "They hear constantly of the great social maladjustment, but no way is provided for them to change it, and their uselessness hangs about them heavily."

At twenty-seven, Hamilton was squarely in adulthood. She was enthusiastic and passionate, but just like the people Addams referenced in her speeches, she didn't yet know where to direct her energy.

Addams continued speaking with a forcefulness that would lead her to bang the podium and round out her sentences with a rhetorical staccato.

"Hull House is an experimental effort to aid in the solution of the social and industrial problems which are engendered by the modern conditions of life in a great city," Addams said from the stage.

Hamilton was hypnotized. A settlement house was a perfect receptacle for the idle drive of a young doctor eager to help other people, and one that ensured she'd be constantly busy.

When the speech was over, Hamilton made her way to the stage and waited in line to meet Jane Addams. When they finally spoke, Hamilton said that she wanted to live at Hull House.

Landing a room in Hull House was not as easy as Hamilton hoped. It took Addams several months to have an opening, during which time Hamilton looked for other settlement houses. There weren't many, and others were poor imitations of Hull House. At one, Chicago Commons on Union Street, a resident told her that she would be of little use if she could only be available in the evenings.

But she was undeterred. "I was resolved that in some way, somehow, I would make my way into a settlement," she later wrote. "I had a conviction that professional work, teaching pathology, and carrying on research would never satisfy me. I must make myself a life full of human interest."

When Addams wrote to offer her a room at Hull House in September 1897, Hamilton dropped by the house that day to accept. She moved in within a week.

When she wrote to her sisters and friends with her exciting news, she had a clear-eyed explanation for what she intended to do. "It was not in a spirit of youthful rebellion against things as they were that I went to Jane Addams' Hull House," Hamilton wrote later. "I had begun to realize how narrow had been my education, how sheltered my life. I wanted to go into that underworld and see for myself."

Settlement life operated on a loose philosophy of being joyfully available and morally helpful. But the house itself operated on a strict set of rules. All twenty-five residents of Hull House had to pay their own way—an inflexible three dollars per week. They needed reliable incomes, and if they had none, they were required to commit to service in the house on evenings and weekends. Visitors had to be approved in advance, and time away was to be kept to a minimum. The hope was that strict rules would attract disciplined people, and mostly they did. A frequent visitor to Hull House later remarked, "These people seldom sing, almost never dance, and their attitude toward each other carries no sign of their sex life."

Meanwhile, grandiose visions of settlement work crashed hard into the reality of the bleak surroundings. The area around Hull House was a dense part of Chicago's Nineteenth Ward, where entire families lived in small rear tenements and wooden shanties. Disease spread easily. Tuberculosis rates were high, as was child mortality. Working with immigrants sounded nice to wide-eyed volunteers, but communication was difficult with people who primarily spoke Italian, Greek, Polish, Yiddish, or German. The garbage went uncollected for weeks and piled up on the sidewalks. People and rats had grown accustomed to each other. Many families lived on less than five dollars a week, a sum that barely kept them warm, let alone clothed and housed. A man's child or his wife getting a meager job in a sweatshop was often the difference between the family's subsistence and starvation.

It usually took several weeks for the privileged people who lived at Hull House to adapt to life in the Nineteenth Ward. Within the walls, Hull House sought to be a utopia of multiculturalism. But outside was not so easily fixable with empathy and enthusiasm. Hamilton and her fellow residents regularly witnessed drunkenness, child abuse, disease, and murder.

The shell shock would disappear when a new resident was called to help, often in an emergency. Jane Addams herself, a woman with deep compassion but hardly any medical experience, had several times been called in the middle of the night to help deliver a baby. Other times, self-reward would come from helping an immigrant learn to read or fill out forms. Many immigrants wanted to earn money, but they also seemed to crave acceptance as *American* with all its rights and privileges.

Hamilton could see this every day without even stepping outside. At night, while she sat in the downstairs parlor, she could observe the desperation of her neighbors.

"How old are you?" she overheard a woman ask an old Irishman who had come for help filling out a job application.

"Sixty," the man said.

"No he isn't," his wife interrupted. "He's forty-five."

"I'm sixty," he insisted, wanting to appear more virtuous. "They should think I fought at Balaclava," he said, referring to the Crimean battle of 1854.

"Very well, and when's your birthday?"

"Say July fourth," the wife said.

"July fourth," he said.

"Do you have your naturalization papers?"

"No, I lost them in the snow in Utah when Brigham Young was dedicating the temple." The man, like many other residents, seemed to cling to historic moments to foster a role, however contrived, in the story of America.

Life at Hull House was communal from sunrise to long past bedtime. The drawing room and white marble fireplaces reminded Hamilton of her grandfather's mansion in Fort

Wayne. Living in a house with constant activity outside one's bedroom was also reminiscent of her childhood under the same roof with more than two dozen relatives.

But in other ways, her comfortable upper-class life in Indiana bore little resemblance to her life of service in Chicago. For one thing, her father's belief in free enterprise and individual responsibility to improve one's life was at odds with a social establishment like Hull House that operated on the premise that marginalized immigrants and racial minorities lacked the economic tools and societal support to simply lift themselves up.

After she moved in, the intensity of the house—and of Addams specifically—intimidated Hamilton. She was twenty-eight and a certified doctor, but she felt out of place. "Miss Addams still rattles me, indeed more so all the time, and I am at my very worst with her. I really am quite school-girly in my relations with her," she wrote to her cousin Agnes in October 1897. "I have pangs of idiotic jealousy toward the residents whom she is intimate with."

In her second week, Hamilton got her first job: door greeter. She was instructed to stand in the foyer and make sure whoever came inside knew where to find food, bathrooms, or one of the many classes or clubs.

After several months as a door greeter, she moved up to giving tours of the house. And from there, she took three weekly shifts in the well-baby clinic, where she gave babies baths. This was simple work for a doctor, but it was harder than she expected. It took time for her to learn not to give advice but to make vague insinuations to immigrant mothers, most of them Italian, whose children suffered from a weakening bone condition called rickets. Many of the women were superstitious about water, and at the first sign of cold weather they sewed their babies into their winter clothes to prevent any washing. Hamilton learned she could get around this fear of water—which was rooted in the fear of unknown illnesses—by following a bath with an alcohol rub and a dab of olive oil on the forehead.

There were genuine reasons to be wary of medicine in the final year of the nineteenth century. Medical knowledge in 1899 was more advanced than ever before in human history, but it still promoted cures as effective as snake oil. Sanitation remained a challenge with overcrowded houses and inadequate sewage systems. Antibiotics did not exist, and many medical procedures, including childbirth, often ended in death. Anesthesia and pharmaceuticals were still in their infancy; most existing drugs were derived from plants and minerals. That same year, German chemists released a general pain reliever called "aspirin," but it wasn't available in the United States yet.

Hamilton had endless opportunities to be humbled by Italian folklore and wives' tales that frequently worked better than anything she had learned in medical school. After she told a group of mothers that breast milk was the only suitable food for infants, according to the established medical norm at the time, one woman brought her one-year-old child to the clinic to prove Hamilton wrong.

"I gave him the breast and there was plenty of milk," the woman said. "But he cried all the time. Then one day I was frying eggs and just to make him stop I gave him one and it went fine. The next day I was making cupcakes and as soon as they were cool I gave him one, and after that I gave him just whatever we had and he got fat and didn't cry anymore." Hamilton could only shrug sheepishly. It was a valuable lesson that highbrow medicine was often at odds with real-life experience.

Trust was a rare commodity among immigrants. Any effort Hamilton made to persuade people in the community that they were valuable and welcome in America could be undermined by frequent episodes of injustice. One day while she was wandering the neighborhood, two Italian workmen were sitting on wooden garbage boxes in front of their tenement. A police officer who was Polish happened to be passing by and told the men to move along. They refused, saying that they were sitting on their own garbage boxes in front of their own house. The officer drew his revolver and shot both men in the chest. Hamilton arrived on the scene moments later when a mob of angry Italians had surrounded the officer and his partner in a nearby house. The neighbors took up a collection and raised $400 in nickels and dimes—from people who hardly had enough to eat—for a lawyer to lead a prosecution effort. But after several months, the police chief said that it was best to move on, and the officer was not punished.

Once Hamilton learned to anticipate and even compartmentalize traumas like this, life provided great opportunities for fun. She loved her time with children and frequently took on unpaid shifts babysitting or entertaining toddlers with games and stories. On the weekends, she'd sometimes walk around the neighborhood and scoop up as many as thirty kids from the stoops of tenements and take them on excursions to the park or to Lake Michigan to feed stale bread to the birds. The teenagers seemed to be especially fond of Hamilton, a figure old enough to admire but not as old as their mothers. For a woman living life on an unconventional path, Hamilton felt she had found her place.

As time went on, Hamilton's medical training made her too valuable to Hull House to simply give rudimentary infant baths and lead the youth brigade around town. As she passed her thirtieth birthday at Hull House in 1899, her responsibilities steadily grew.

It started when a woman knocked on Hamilton's door and

asked if she could make a house call to diagnose a wheezing baby.

When she arrived, Hamilton calmed the infant and offered a range of medicines. It was common at the time to prescribe sick children a newly released drug called "diamorphine," created by an English chemist. It worked especially well for children's coughs, sore throats, and colds. It also seemed to create a deep dependency in children, so much that the same German company that developed aspirin started to market the new drug under a name derived from the German word for "heroic" and called it "Heroin."

While she inspected the baby, she was distracted by a man coughing in the next room.

The incessant coughing sounded like it scraped the bottom of his lungs. Hamilton asked the mother if the man was okay.

The woman said he had been coughing like that for weeks.

Once the baby was settled, Hamilton suggested she have a look at the man, curious if he had an infection. Between coughs, he said he was feeling fine. But his wife interjected that several of the man's friends had the same cough. They all worked at a nearby stockyard, and none of them had worked there longer than a few months.

This sort of thing began to happen more. While treating children, Hamilton would be pulled into cases of their parents, usually their fathers, who seemed to all have strange similarities. She noticed that men who worked in stockyards near animal pens had high rates of pneumonia and rheumatism. Men who worked in steel mills seemed to have a peculiar brain fogginess caused by breathing carbon monoxide that doctors would later call "dementia."

But none of the men seemed to have it as bad as those who worked as painters or in chemical factories. Those men Hamilton could spot simply by passing them on the street. They were pale, thin, and prematurely wrinkled at thirty years old.

If Hamilton could get them to talk to her, they complained of indigestion and searing pain in their wrists. Several said that after barely a few weeks on the job, their hands had gone completely limp. For this they were fired and told they were clearly unsuited to hard work. These sorts of things had gone on for years, the men told her. Some of their fathers had died from the very same symptoms. Nothing ever changed, and nothing indicated it ever would.

Hamilton refused to accept such a reality. The rates of sickness bothered her, but the resignation was even worse. Where other doctors might have shrugged and turned away, Hamilton felt a moral fire within her. Everything in her life— her upbringing, her medical education, and her social work at Hull House—prepared her for this work.

She had no master plan to solve these problems, but she knew how to start. She kept a journal and recorded where men worked in one column and their symptoms in another. Her handwriting was barely legible, but the purpose was only to communicate with herself.

Not all men were getting sick, she learned, but the worst cases seemed to be concentrated together. And within a few weeks, her chicken scratches seemed to point to an undeniable finding. The cases of wrist palsy that sickened men and caused painful deaths all seemed to be coming from one factory, and it was one that made lead.

# 4

## Chicago, Illinois — 1909

Hamilton was barely thirty-one, but she was encountering an enemy that had been familiar to doctors for centuries. At number 82 on the periodic table, lead was a remarkable metal that stood out for its unique set of properties: It was easy to find, abundant in nature, and soft to mold. It held its pigments well, resisted corrosion, and even tasted sweet, which made it almost miraculous to behold, marvel at, and even eat.

The oldest sign of humans interacting with lead is a necklace roughly 8,000 years old that was found buried in a cave in Turkey. Lead was seemingly used daily, and the ancient Egyptians were the first to use particles of it in foods, medicines, and cosmetics. Lead weights helped sink their fishing lines and coat their pottery.

The Egyptian appreciation for lead eventually spread to ancient Greece, where the philosopher Hippocrates was one of the first people to wonder if lead, the miracle metal in any situation, might also have a dark side. He noticed that people who ate rich diets of food and wine—particularly when stored in lead containers or improved with lead preservatives—came down with gout. But no one could prove it. A man was lucky in his era to make it to forty. If lead didn't kill him, a long list of other maladies—like bubonic plague, influenza, measles, typhoid, typhus, smallpox, fire, blunt trauma, homicide, cancer, venereal disease, alcohol, or opium—might get him first.

Any hesitation about lead was buried by the time of the Roman Empire, which considered lead the everything element—the plastic of its era. Lead made pots and cooking utensils last longer. Lead helped urns resist bacteria that would turn wine to vinegar. And it made pigments in makeup brighter on people's faces. Its sugary flavor made it an easy additive in dishes, and one that could preserve food for weeks. Governors used lead to line the routes of aqueducts, and when it came time to install water pipes into homes, the natural thing for the Romans to call the new system was "plumbing," from the Latin word for lead, *plumbum*.

Lead accumulated in people's bodies at a dizzying pace. In modern times, most people take in about 0.3 milligrams per day from trace amounts in food, water, and the air. The Romans ingested as much as 250 milligrams. Fortunately for them, a portion of lead is expelled from the body, unlike a more corrosive element like mercury, which accumulates quickly. But a fraction of lead stays behind, a slow-motion saboteur of the body's bones and organs.

Lead was believed to draw out a person's insanity the way getting drunk draws out inhibitions. But not everyone went insane in Rome. It was mostly rich patricians who had advanced cooking tools and large houses with lead pipes. The erratic behavior of emperors like Claudius, who was known to frequently drool and have fits of excessive and inappropriate laughter, and Caligula, who once tried to appoint a horse as a governing consul, make more sense when considering that the average high-class Roman consumed ten times more lead than someone in the lower classes. Historians have long debated this point, but there remains an unmistakable coincidence in timing between the strange paranoia, mania, and physical immobility associated with lead poisoning and the accelerated collapse of the Roman Empire.

Without an advanced civilization to mine it, industrial uses of lead subsided for hundreds of years. The next groups to discover lead's abundant benefits were the people of India and China, who seemed to realize anew that lead could preserve food, coat pipes, and brighten the colors of face powders. Bronze was stronger and less brittle—enough to have a full "age" named after it—but lead was a simple and affordable standby, especially for the lower classes. It was here that lead began its expansion from a metal that mostly poisoned the rich to one that also sickened people too poor, illiterate, or unlucky to know of its dangers.

Lead eventually circled the world. Along with their diseases and violence, European colonists began to introduce large quantities of lead to colonies in Africa, North America, and Australia. Settlers in Virginia started mining for lead almost immediately upon the founding of Jamestown. In the Congo and Australia, indigenous people who had never heard of lead started pulling it out of the ground as fast as they could to trade for better commodities with the richly endowed British.

In 1900, Hamilton's second year at Hull House, America became the top miner of lead in the world. The country's booming population demanded the construction of more houses, the best of them built with durable lead pipes. The uses of lead were almost limitless. An ad for Dutch Boy paint, the era's fashionable paint brand, declared that homes painted with lead paint would have "color harmony." Another marketed paint colors for children's rooms with delectable names

that suggested kids could lick the walls. Lead was the enlight-
ened solution to a healthier and more prosperous life, like
breathing fresh air. People with drab walls coated in old-style
paint began to feel the pull of their friends and neighbors to
upgrade to the new stuff. The powerful marketing machine
of the lead industry branded lead "the useful metal."

The best part of lead, and the quality that kept it relevant
and exciting rather than toxic and dangerous, was the way
it corrupted the body slowly. Unlike lead *poisoning*, which
came quickly and clearly from consuming too much lead all
at once—perhaps through a lead bullet wound, for example—a
lower level of lead *exposure* was so difficult to observe that
a person wouldn't even be aware of it. Doctors in the early
twentieth century had no idea exactly how much lead would
cause a strong reaction. It was thought that one could recover
from a low amount of lead with no long-term effects—similar
to drinking alcohol—but knowing the threshold where exposure turned dangerous was hard to determine, and different
for every person.

This was why Hamilton was so indignant. She could
clearly see where the sick men in Chicago seemed to be
getting sick. But there was almost no concern from the
men's bosses or any government leaders about how to stop
it. A man could erupt in anger, fall into hysterics, or even
drop dead on a factory floor. But lacking any standard for
how much lead exposure was too much, and whether the
man in question had crossed that threshold, the boss could
simply blame the man's fatigue or his poor work ethic—
usually in reference to an ethnic stereotype—and leave it at
that. Putting the blame on the worker himself also conveniently absolved the factory owner of any moral or financial
obligation.

Hamilton grew accustomed to these tales of naked injustice. "Living in a working-class quarter, coming in contact

with laborers and their wives, I could not fail to hear tales of the dangers that workingmen faced, of cases of carbon-monoxide gassing in the great steel mills, or painters disabled by lead palsy, or pneumonia and rheumatism among the men in the stockyards," she later wrote in her memoirs.

The stories gnawed at her deep-seated sense of empathy, and she was unwilling to sit by and do nothing.

Hull House was the perfect place to channel Hamilton's outrage. As it happened, the arrival of the twentieth century opened a new era for the settlement. It had grown from a scrappy service project to a venue of high-minded ambition where social change began. It was written up in newspapers and magazines around the world, and famous dignitaries and even celebrities often stopped by to see for themselves the economic melting pot Addams had created. Teddy Roosevelt paid a visit. So did the future "king of swing" Benny Goodman and the Irish novelist Francis Hackett. Writers and scholars from Europe made visits to see a working experiment in American diversity. Economists who had once mocked Addams's idea for a settlement in the middle of a slum arrived weekly to marvel at how well it worked.

In greater numbers than the dignitaries, however, were the young people who arrived daily holding nothing but knapsacks and a starry-eyed sense of aimlessness. In its third decade, Hull House had become a place for students to go after college or graduate school to meet other young people who shared their goals in writing, business, or city planning but had little idea how to start their careers. If a room was available, they'd be put to work teaching English class or assigned to neighbors who were floundering under illness or poverty.

Many of them, like Hamilton, were immediately horrified

by deep disparities in the social order. They were dispatched to collect signatures for political campaigns and to petition authorities for more public playgrounds and recreation centers. Those with science or medical experience like Hamilton conducted informal studies on things like the reading habits of children, how much cocaine teenagers were consuming, and the causes of school truancy.

As the months turned to years, Hamilton's notebooks filled with the stories of hundreds of fathers and husbands who succumbed to either debilitating illness or death that left their families on the brink of survival. As more time went on, her sense of purpose seemed to fray: She was extremely effective as a doctor treating the men, but she was also powerless to address the root causes making them sick.

Until two incidents occurred that made Hamilton's polite persistence boil over into fiery outrage . . .

The first incident happened in 1906 when a group of workmen were sent out to repair one of Chicago's pumping stations in Lake Michigan. The station sat on a small floating island, barely large enough for ten men to stand on. About an hour after a boat dropped the men off, an electrical fire broke out. With the entire island ablaze, the men had no choice but to jump into the deep lake. More than half of them drowned before the tug came at the end of the day to pick them up.

The accident didn't warrant even a passing mention in the Chicago newspapers. It was only because of a young journalist named William Hard, who lived at Hull House, that the preventable accident got any attention at all. Hard was a progressive journalist known to business bosses and government officials as a "muckraker"—a person who wrote stories that riled up the public. Hard published an article in *Everybody's Magazine* about how the company that employed the men did nothing except pay for the victims' funeral expenses. (Hard pointed out that many people considered the funeral payments "generous" since the company had no legal obligation to pay anything at all.) But no one seemed to care that the accident had long and costly implications for the dead men's wives and children, who received nothing, not even a basic acknowledgment that they had lost a breadwinning family member and would fall into poverty.

The second incident shocked Hamilton even more. There had been an accident at one of the steelworks on the east side of Chicago. A group of workers were injured and taken to the company hospital, a drab facility that existed solely so the company could hide its mishaps. The men were isolated and permitted no visitors, even their families. After several days, the wives of the men grew so desperate for any information, including how badly their husbands were hurt, whether they would recover, or whether they were even still alive, that they came to Hull House for help. Hamilton tried for days to get answers, but the corporate office ignored her too. Even worse than the poor conditions, she believed, was the arrogance of a company that devalued human life to protect itself.

These two incidents horrified Hamilton because they occurred in front of her eyes. But there were hundreds of similar scenarios around Chicago—and all over the country—of men being crushed by the industrial machine. If their misfortune got scant attention, the misery of their wives and children got even less. *Their* plights often occurred in slow motion, a steady grind from birth to death, especially for the women who worked in sweatshops around South Halsted Street. The women were forced to arrive before dawn and work without breaks until their fingers were raw. Then they would carry bundles of men's and boys' T-shirts and trousers to their tenements to finish at night. Often their children would help late into the night by pulling out basting threads. The women could hardly speak English, which ruled out any attempt at unionizing or advocating for better wages. The number of hours a woman worked was entirely decided by how much money she needed, which led some to work as many as twenty hours a day.

Hard, the journalist, grew tired of the sheer volume of these realities for workers, and over time his articles transformed from objective news reports into angry screeds of activism. Why was no amount of horrifying detail ever enough to elicit even a crumb of compassion from business bosses? Why was the general public so dismissive of the burdens of the poor? Hard fixated on the sweeping scale of the problem, of how many men and families were disrupted. "Every year the stream of industrial accidents flows on," he wrote, "and every year it sweeps hundreds and thousands of families away from their little perilous stations of self-respecting independence down the irresistible current first to poverty and then to charity."

By 1907, Hamilton had risen in seniority at Hull House. She found younger people to do her former work of making house calls and running the baby clinic. Now, at thirty-eight, she focused her attention on why American workers were so poorly treated and no one seemed to care. She gave regular soapbox lectures around the dinner table. She wrote letters and articles for the *Chicago Tribune*. Someone gave her a book—*Dangerous Trades* by a British doctor named Thomas Oliver—and as she read, she marveled at how Oliver discussed in clear prose "the dangers to life and health to which trade workers were liable."

Hamilton finished Oliver's book in one night and then went to the library to find more on the subject. She found books about industrial diseases in Dutch, Italian, and Spanish, and, thanks to her father's early insistence on learning languages, she read most of those too. "In those countries industrial medicine was a recognized branch of the medical sciences," she wrote. "In my own country it did not exist." The only explanation, she surmised from conversations with friends, was that her desire to protect workers was somehow at odds with the pillars of capitalism, in which drudgery was supposed to motivate workers to work harder.

Hamilton noticed that the shortcomings in basic worker safety in America, as compared to other countries, were excused by blithe arguments that America's factories were cleaner and more efficient, with workers who were better trained than anywhere in the world.

By 1909, Hamilton had transformed from the wide-eyed young doctor who had come to Hull House in 1897 into an impassioned doctor who was furious at seeing the sorrows of injustice play out every day. The man who lost his hands. The wife who lost her husband. The child born with one eye or a stump for an arm.

Her complaints became frustrated tirades, usually while commiserating with another doctor at the University of Chicago about the complete disregard for even basic precautions. What was the point of knowing for thousands of years that lead was a devastating poison if men were still sent daily to refine the metal, prepare it for use, and breathe its dust?

One day she ran into Charles Richmond Henderson, a professor of sociology at the University of Chicago, who had an idea. Henderson had recently returned from Germany, where he had studied the various things making German workers sick. The study wasn't for the sake of the workers; workers could never organize such a study, let alone fund it. It was ordered by a group of German health insurance companies, the *Krankenkassen*, that wanted to know how they could incur fewer payouts to sick workers. A form of capitalism, in other words, that worked *in favor of* the workers. Henderson spent months inspecting German factories and reported his findings with basic recommendations, many of which were quickly adopted, including increasing factory ventilation and laundering worker uniforms once a week.

Why, Henderson wondered to Hamilton, shouldn't a similar study be made in Illinois? And if no industry would fund it, then the government should.

It helped that Henderson knew the governor of Illinois, Charles Deneen, a progressive leader who had stood up for the rights of immigrants and signed a 1905 law criminalizing lynching in Illinois. Deneen agreed to appoint a panel he called the Deneen Commission that would be tasked with "thoroughly investigat[ing] causes and conditions relating to diseases of occupation."

Knowing her level of passion for the subject, Henderson got Hamilton a seat on the commission. A few other doctors were recruited along with government officials. The group was to find answers and make clear and actionable recommendations. Starting on New Year's Day in 1910, they would have $15,000 and exactly nine months.

At the first meeting of the commission in January 1910, Hamilton and the eight men discussed exactly what they should study. Did "industrial diseases" mean only debilitating conditions like men working with poisonous materials? Or should their study be broader, looking at the punishing conditions that, given enough time, would lead to chronic stress and the body's breakdown?

"We were staggered by the complexity of the problem we faced and we soon decided to limit our field almost entirely to the occupational poisons," she wrote, "for at least we knew what their action was." Time was limited, so the commission agreed on twenty-nine known poisons.

1. Ammonia
2. Aniline
3. Antimony
4. Arsenic
5. Arseniuretted
   hydrogen gas
6. Benzene
7. Carbon dioxide
8. Carbon disulfide
9. Carbon monoxide
10. Chloride of lime
11. Chlorine
12. Chromates
13. Dinitrobenzene
14. Formaldehyde
15. Hydrochloric acid
16. Hydrocyanic acid
17. Hydrofluoric acid
18. Lead
19. Nitrobenzene
20. Manganese
21. Mercury
22. Methyl alcohol
23. Nitrous gases
24. Phosphorus
25. Picric acid
26. Pyridine
27. Sulfur chloride
28. Sulfuretted hydrogen
29. Sulfurous acid

Hamilton volunteered to steer the investigation on lead. It was the most widely used poison on the list and also, she believed, the most destructive. She learned in medical school about the difference between lead poisoning and lead exposure. Lead poisoning would cause excruciating attacks that included convulsions, encephalopathy of the brain, and temporary blindness. But even small doses of lead exposure, which appeared to have no immediate effect, would advance into longer-term conditions like wrist-drop and "premature senility" when a forty-year-old man could look twice his age.

Hamilton was eager to begin but didn't know where to start. Investigating lead was like "trying to make one's way through a jungle and not even being able to find an opening," she remarked. The first step was to determine which industries in Illinois used lead and which companies used the most. She decided she would get to know factory leaders and inspectors while simultaneously asking doctors and pharmacists if they could recall clusters of lead-related illness.

After a month she had a list of three hundred factories and smelters and about seventy different industrial uses for lead. The most destructive among them appeared to be those making can seals, laying electrical cables, and polishing cut glass. From there, she tried to identify the factories that had the most revenue in Illinois, which suggested the most pressure on workers. And then she started visiting them.

On March 16, 1910, Alice Hamilton arrived without an appointment to the National Lead Company's Sangamon Street works on the east side of Chicago to take a look around. The smelting plant made white lead and lead oxide for use in paint pigments, anti-corrosion metal, and automobile batteries.

After she introduced herself, the receptionist called one of the vice presidents to come meet Hamilton. Just as she expected, the vice president, a man named Edward Cornish, was bemused to see her. She told him men were getting sick in his factory and that the governor had appointed her to investigate.

Cornish laughed her off. His plant was immaculate. "A model factory," he said. He went to the door and shouted to a passing workman to come in.

"Did the lead ever make you sick?" he asked the worker.

The man stammered, "No, no, never sick." He had visible scars on his face.

"Are any other men sick?" demanded Cornish.

"No, no, all good."

He turned back to Hamilton. "There, you see!"

Hamilton said no. She was polite but firm. "Your men are breathing white-lead dust and red lead and litharge and the fumes from the oxide furnaces," she said. These were poisons, she pointed out, and they affected Cornish's men the same way they would anyone else.

She asked again to look around.

Cornish relented. He had nothing to hide, he said, so he let her in.

She walked by the kettles and furnaces. She observed the sweating men holding shovels and prods. The air was stifling, full of humidity, and several times she covered her nose and mouth to fend off the head-splitting odor wafting through the unventilated rooms.

On the floor were pieces of scrap metal and a film of metallic dust, some of which also floated visibly in the air. A few men wore dirty uniforms, but most wore the same stained overalls to and from home every day.

Those who worked in the most noxious areas, including the internal windowless melting room, were Black. But even the white workers had little protection from the dust. A pile of rubber respirators sat in one corner, none of them used. The men seemed to find them onerous and unfashionable, and instead tied dirty handkerchiefs around their mouths.

Over the next few months, Hamilton inspected more than twenty facilities. She performed the same dance with bosses who feigned shock at the suggestion that their employees were treated with anything but the utmost generosity. But the men told a different tale, if they were willing to speak freely. For those who weren't, Hamilton found ways to visit them in their homes at night, where she'd interview them for hours on the condition she wouldn't use their names in her report. One man told her about a young immigrant from Bulgaria who'd worked on one of the production lines but seemed to have gone crazy a few weeks prior; he was removed from the plant in a straitjacket and later died. She heard of another man, also an immigrant, who was put to work making paste for batteries and moistened his fingers on his tongue. He lasted ten days. A foreman told Hamilton that most of his men worked only a few weeks before they got sick. Almost all of them suffered the same symptoms, starting with hallucinations that made them see butterflies. There was no payment for being sick, nor did there appear to be much effort from managers to keep the men healthy. Almost no facility had sinks where men could clean their hands before returning home. But the men uniformly reported they were grateful for the job and the few dollars they made each day. And they all asked repeatedly that their complaints not get back to their bosses and jeopardize their work. Even a job that made them sick was better than no job at all.

Several weeks after Hamilton finished her work, the commission published a report of its findings.

It was, the group concluded, the "moral duty of every civilized state" to promote the health and safety of its most helpless people, especially ones trapped or forced to work in menial jobs unseen by the broader public.

Several of the bosses whom Hamilton had visited incorporated the changes she suggested. The report reflected poorly on the companies and on the egos of the men who ran them, who relented to modest changes that would make them appear virtuous. Cornish, the vice president from the Sangamon Street plant, was especially energized. After Hamilton showed him proof that twenty-two of his workers had cases of lead poisoning that required hospitalization, he went on to reform every plant that the company operated in Chicago. That meant installing large exhaust systems to capture the lead dust and providing better uniforms and handwashing stations. He also created a medical department in the plant to monitor the workers and observe symptoms of possible poisoning. He vowed that he would urge plant officials in other states to make the same changes.

The final report first went to Governor Deneen and was then sent around to Illinois lawmakers and the Chicago media. When stories started appearing, news of the investigation seemed to genuinely shock the public. For the first time, definitive statistics and detailed case reports provided a close-up portrait of the epidemic of poor working conditions—and not one that could be easily blamed on an inferior race or ethnicity. Not only did it undermine the country's spirit of capitalistic dominance and American exceptionalism; it also showed that apathetic bosses and lawmakers were actually costing America jobs, energy, and efficiency.

Within a few months of the commission's report, which included Hamilton's deep investigation of lead, the Illinois legislature passed a law that flipped a switch in favor of workers almost overnight. Bosses were required to compensate workers who caught diseases from poisonous fumes, gases, and dust. Companies were required to buy insurance against such cases, and once insurance companies were involved, they provided better enforcement than the government ever could. Insurers sent inspectors to visit factories to make sure that the causes of any illnesses or accidents were minimized. Floor managers gave workers better uniforms and required them to wear face masks. Factory foremen opened windows and installed fans to blow out toxic air. In some industries, like textiles, workers received more breaks and modest increases in pay. In others, like lead processing, foremen earned small bonuses if they found ways to reduce the amount of lead dust in the air. The costs of workplace danger were shifting ever so slightly from workers to employers, proving that capitalism—when paired with modest legal requirements—could work for more people than just those who were already rich.

Hamilton felt like she had triumphed. She believed that progress came in inches, not miles, and that one success, however small, would smooth the path to the next one. On June 10, 1911, after Governor Deneen signed Illinois's workers' compensation bill into law, Hamilton celebrated at Hull House.

Jane Addams made a toast, declaring that Hamilton's work proved settlement work was productive and worthwhile.

But it wasn't enough. It would take three decades for a majority of U.S. states to pass similar laws. And some of the states that did pass regulations required only infinitesimal advances designed to calm public outrage without losing the favor of big businesses and their generous campaign dollars. In Pennsylvania, compensation for sick workers was extremely low, and immigrant workers who were not full citizens did not qualify to make claims.

What was more, while the worst conditions earned new attention, there were still more commonplace burdens like long hours, low wages, and racial discrimination that affected millions more workers every day, and those would be much harder to address.

Exactly eight weeks after Hamilton's commission report came out, the nation would get a grim reminder of the routine cruelty still forced on America's lowliest workers.

On Saturday, March 25, 1911, just after four in the afternoon, a worker in a garment factory in New York City tossed a cigarette in a waste bin. Within thirty minutes, forty-nine workers were suffocated by smoke, thirty-six burned up in an elevator shaft, and fifty-eight jumped from the top floors of the ten-story building. To prevent workers from taking unauthorized breaks, the doors to the factory had been locked.

# Part Two
# A PERIODIC QUEST

# 5

*Midgley testing chemical samples in his*
*laboratory engine indicator*

## Dayton, Ohio — 1916

Gasoline didn't make the cut for Hamilton's study of poisons. In 1911, no one was drinking gasoline or bathing in it. The worst exposure came from catching a whiff at a filling station while a person waited to get their windows washed or oil changed.

In fact, gasoline, and the raw crude oil from which it originated, was so far from poison status that it was building its reputation as a miracle gift from the planet, especially in places like Ohio. Gasoline was exciting. It was literal potential energy for the steeply rising number of American cars—3 million and counting in 1916—that fueled the fast-growing American economy. People didn't fear gasoline. They wanted more of it.

As a substance, the fuel that powered the internal combustion engine was barely half a century old. Almost every culture in history had taken a liking to raw petroleum. The sludgy black gunk that seeped out of holes and wells appeared to be good at heating, waterproofing canoes, reducing fevers as a topical ointment, and even, in the case of George Washington's troops, treating frostbite and battle wounds during the Revolutionary War. But it wasn't until 1846 that a Canadian chemist named Abraham Gesner started experimenting with refining a form of raw petroleum known as bitumen. He called his new fuel kerosene, from the Greek words for "wax oil," and demonstrated that when it was used to light lamps, it produced a brighter and cleaner flame than conventional coal or whale oil. Gesner, who was proud that he was saving whales from slaughter yet was completely oblivious to the long-term effects of fossil fuels, considered himself an early environmentalist.

Petroleum became a valuable substance in the 1850s thanks to a series of pioneering men who founded companies on the idea that they could drill for petroleum in the rocks of Pennsylvania and Connecticut. For a time, there was more oil than anyone knew what to do with. Uncorking the pressure on a buried well would cause the oil to squirt up from the ground faster than it could be contained, which immediately created crews of well-paid laborers to build barrels as fast as possible. Mismatched barrels with wildly different weights made the oil hard to move, so after a few years barrel sizes were standardized at exactly forty-two gallons, or enough to weigh three hundred pounds, the maximum weight that two men could carry.

Gasoline didn't appear until 1865, the same year the Civil War ended. Its first use was to provide light in large buildings, like factories and grain mills, for which kerosene wasn't powerful enough. When the first motorized wagons arrived in the 1880s, gasoline grew in popularity, since it contained enough force to power an engine. By then one company had come to dominate the making of all petroleum products, and it would profit handsomely from what would become the world's insatiable appetite for gasoline. That company, named Standard Oil for its "reliable standards of quality and service," was run by the ruthless capitalist John D. Rockefeller. But Rockefeller's company quickly became known for its low standards. It devoted its considerable resources to bankrupting rivals, dominating the market, and making as much money as possible. It invested almost nothing in improvements in cleanliness and efficiency that might have demonstrated to future oil companies how to operate sustainably. Rockefeller, who became the world's first billionaire, gave birth not only to the country's first monopoly but also to the blueprint of the great American corporation.

The success of gasoline made possible the success of gasoline-powered automobiles. Steam could make a car go fast: In one demonstration in 1906, a steam-powered car topped 126 miles per hour. But steam required enormous amounts of fuel to heat the water, and it had to be burned outside the engine, which was inefficient. The miracle of a fuel like gasoline was that it could be burned *inside* the engine— internal combustion, in other words, which promised to reduce the cost of operating an automobile and maximize an engine's performance.

Production of the first commercial automobiles began in the United States in 1896, the year before Alice Hamilton

moved into Hull House and five years before Midgley experi-
mented with tree sap on baseballs. Previously, the dominant
way to get around a city was to walk or, for the lucky ones,
to connect a horse to a carriage with wheels. Horse-drawn
carriages had been the standard of transportation for centuries,
with only modest innovation. Wheels got sturdier, carriages
got steadier, and some brilliant person lost to history thought
to install springs under the seats to cushion every bump.

With advances in motors and engines, the transition to
automobiles started slowly—so slowly that, if not for the noise
of an engine, a person might not even notice that a carriage
passing by didn't actually have a horse pulling it along. Once
the design changed to include an indoor cab that sat lower to
the ground, the technology started to appear not as a modest offshoot of a horse-drawn carriage but as a new contraption entirely. These marvels were so novel that no one knew
what to call them. Someone suggested "autotruck"; another
proposed "autowain," from an old Saxon term for wagon.
Neither was acceptable to the *New York Times*, the newspaper
of authority, whose editors seemed to be extremely grumpy
about a technology disrupting America's streets. "There is
something uncanny about these new-fangled vehicles," the
paper wrote in a cantankerous editorial on January 3, 1899.
"They are all unutterably ugly and never a one of them has
been provided with a good or even an endurable name." The
paper pointed out that the French had begun using the word
*automobile*—half Greek and half Latin—but such a sloppy
mishmash of languages, wrote the paper, "is so near to in-
decent that we print it with hesitation." The paper suggested
that horses were taking revenge on ungrateful humans by
"stumping us to find a respectable name for our noisy and
odorous machines."

To the chagrin of the *Times*, "automobile" stuck. Over the
next decade, anyone even casually wandering the streets of

America's big cities noticed a boom in automobiles on the roads. Or, as they later became known in slang, "cars," from the old Latin word for a two-wheeled cart. At first, cars were a shocking spectacle that seemed to defy the boundaries of imagination. Towns once accustomed to the rhythmic clip-clop of horses suddenly roared with the loud purr of engines. The novelty wore off, but it was easy to see a profound shift in society. Young people flocked to the new technology, their parents warmed up more slowly, and older people lamented a world that was changing too fast.

The most popular cars were ones made by Ransom E. Olds, who had started the Olds Motor Vehicle Company in 1897. By 1904, Olds was selling nearly 5,000 vehicles a year, an astronomical number in an industry less than a decade old. Every one of Olds's automobiles was handmade, which made them absurdly expensive and accessible only to the idle rich. Olds's cars sold for $650, just over the yearly salary of the average worker. Everyone else could only watch cars and their glamorous occupants pass by. Although glamour was debatable. Olds's vehicles and other early automobiles were dirty. They spewed clouds of black smoke, and if drivers and passengers didn't get engulfed in the exhaust, they would be splattered with dirt from the road. As time went on, many poorer people resolved that for the high cost and all the hassle, they would rather walk.

If the price came down, though, then it was reasonable to imagine there would be demand from thousands of common people, even millions. This was the philosophy of Henry Ford, a Michigan engineer who had worked under Thomas Edison. At age thirty-three, while Ford was working at the Edison Illuminating Company of Detroit, he invented his first experimental automobile, which he called the Ford Quadricycle, a four-wheeled open wagon with a seat barely big enough to hold two people. He went on to sell three of them.

Two years later, he built another prototype, which also sold poorly. Three years after that, Ford decided to go it alone and start his own company, the Henry Ford Company. But after squabbles with investors, he left that too. Finally, in 1903, he started the Ford Motor Company. He had a vision: He would create an automobile for "the great multitude" that the average person could afford, drive, and maintain.

"It will be large enough for the family, but small enough for the individual to run and care for," Ford wrote of his initial vision. "It will be constructed of the best materials, by the best men to be hired, after the simplest designs that modern engineering can devise. But it will be so low in price that no man making a good salary will be unable to own one." The idea sounded nice. But the feedback Ford heard most was that if he compromised on the laborious and expensive process of making automobiles, he would produce vehicles of such low quality and reliability that he would be out of business in six months.

Energized by the challenge, Ford began experimenting. He started in 1903 with the Model A, an automobile that held five gallons of fuel and had the power equivalent of eight horses. Over the next several years he worked his way through the alphabet with twenty different models. He abandoned the D, E, G, H, and I, and made incremental advances on the B, C, F, N, R, S, and K, including different-sized engines, cylinders, and horsepower. The variety was a big hit for Ford's salesmen, who loved offering different options to different buyers. But too much variety meant too many parts were slightly different sizes, which was bad for production.

Finally, in 1908, Ford landed on the Model T. It was instantly his favorite because it was simple, just four main fragments that fit together like a puzzle, which could be assembled in minutes by even the lowest-trained worker. If one of the panels broke while the car was being driven, it could be replaced quickly and seamlessly, often for less than the price of repairing it. This required that the panels be identical in every way, including their color, which led to Ford's famous remark that a customer could select any color he wanted so long as it was black.

Ford invented the automobile as a commodity, and it began to sell as one. Within days of announcing the Model T, Ford received 15,000 orders. It was still expensive at $825—more expensive than the cheapest models sold by Ransom Olds—but Ford promised that not only could anyone drive and maintain a Model T but also that the long-term investments the Ford Motor Company was making in its manufacturing process would bring the price down, not up. Sure enough, by 1927, Ford was making 10,000 cars every day and selling them for just $325. The Model T became the most sold automobile in the world, a title it held until 1976.

Ford's success did not eliminate his rivals. It encouraged others to copy him. But despite Ford's innovation in manufacturing, the Model T still had several flaws, none more glaring than the onerous, dirty, and dangerous hand crank to start the engine.

As it happened, this was the very problem that allowed Charles Kettering to make his name in the nascent automotive industry. In 1911, while the first Ford Model Ts were rolling off the production line, Kettering put the finishing touches on what he called his "engine starting device." The system relied on a battery to power a small motor that would start the engine—the same job formerly done by a man at a crankshaft. Once it was running, the engine would recharge the starter battery so it could power the motor to start the engine again in the future. It was a perfect loop, so long as the battery didn't sit idle for too long and drain.

The first order for his ignition system came from Cadillac, one of Ford's competitors, which believed that if *its* cars didn't require a laborious hand crank, consumers might be willing to spend a little more. Ford thought they were foolish: Needless contraptions made cars more complicated and more expensive. But on this he was wrong. He held out almost a full decade, until 1919, when he relented and began installing self-starting ignitions on new Model Ts.

Kettering earned his chops with a brilliant invention that moved the automotive field forward. And in time he would be richly rewarded.

As Ford built his empire, a growing automobile company called General Motors figured out a way to compete. Alfred P. Sloan, the savvy CEO of General Motors, sensed that consumers were growing tired and bored of driving the same Ford car. Sloan thought of changing the production timeline of all General Motors vehicles to create a system of "planned obsolescence." If GM cars were given modest improvements every year, customers would feel they needed to upgrade.

To further outfox Ford, Sloan also wanted more streams of revenue for General Motors than only selling cars. He started to acquire companies that made car parts. He bought a refrigeration company later named Frigidaire and invested in a small automobile financing company that he turned into a multimillion-dollar business.

As luck would have it, Sloan's buying spree also scooped up Kettering's company, Delco, which catapulted Kettering from rich to richer.

After the acquisition, Delco was rebranded as the General Motors Research Corporation, which brought new prestige and urgency, especially for Thomas Midgley, who went from working for a modest research company to one of the biggest names in automobiles, all without changing bosses.

It also brought a feeling of limitless opportunity. With Sloan's backing, Kettering's record of success, and Midgley's engineering smarts, it was easy to imagine that the new research division of GM could solve the most vexing problems in automobiles—and maybe even the biggest one.

Midgley settled into life in Dayton in 1916. His two kids, ages four and two, took quickly to the city's parks and the summertime circuses that rolled through town on the New York Central Railroad, while Carrie Midgley enjoyed indulging her artistic curiosity at a new art museum.

Meanwhile, Thomas Midgley loved working. He arrived at the office early and often started the day in Kettering's office, shooting the breeze with his boss about new developments in automobiles and the various models rolling off production lines at different companies. Kettering filled Midgley in on the revenue numbers from sales of self-starters and lighting systems. Then Midgley would wander to the laboratory and fiddle around with a contraption he was working on or putter over to see what a colleague was up to.

Midgley and Kettering had the odd idea one day to try to see *inside* an automobile engine to judge whether it was operating at peak efficiency. Midgley thought up a plan for such an endeavor. He spent two days building a beaverboard enclosure around an engine to shut out light. He took two thin strips of wood, two nails, and a tomato can to fashion a photographing device. Then he attached a piece of photographic paper to the can with rubber bands. While the engine ran, Kettering spun the tomato can on its nail pivots while Midgley operated the shutter. The photograph was a blurry mess. But the two men rejoiced at creating the first photograph of an engine, which would spur them to take better and better photos so they could see for themselves exactly how gasoline was burned.

These types of happenstance bouts of curiosity might have continued if not for the Great War, as it was called, that engulfed the United States in April 1917. Wartime meant a pause to frivolous activities and an all-hands-on-deck mentality across America's industrial belt. Washington directed factories to turn their efforts toward making munitions, food, and equipment. Midgley, who had by then devoted his full attention to fuel combustion, volunteered to try to find a more powerful fuel for American fighter airplanes to match the airpower of the Germans.

Midgley heard a rumor that the Germans were using a flammable kitchen detergent called cyclohexane in their airplanes, so Midgley tried to make the same thing. Cyclohexane had six carbon and twelve hydrogen atoms ($C_6H_{12}$). To mimic it, he thought to start with a similar molecule called benzene, which has six carbon and six hydrogen atoms ($C_6H_6$). If he could add six more hydrogen atoms to benzene, a process known as "hydrogenating," he could make his own jet fuel.

This was the exact sort of experimentation Midgley loved: setting a goal and finding a way to make it work. He gathered the equipment to synthesize and isolate hydrogen atoms and hooked up various test tubes and beakers to run the mixtures.

One day, as Midgley bent over to inspect the tank he was mixing, a plug blew out and shot metallic particles in his eye.

Midgley stiffened in pain. The metal particles were lodged in his cornea.

He tried holding his eye under water for a while and then tried to pick out molecular chunks with a cotton ball. When that didn't work, he called his doctor, who came quickly and suggested several treatments involving warm liquids to soften his cornea. But no luck.

After several days of pain and at risk of blindness, Midgley decided to experiment on himself.

He thought about bathing his eye in pure liquid mercury. Mercury was highly toxic, but he thought mercury's atomic structure would form an alloy with the metal bits and flush them out.

Surprising everyone, it worked. His eye healed completely.

Midgley was so delighted with his success that he submitted an article about the incident to an elite journal run by the American Chemical Society.

As it happened, the eye incident seemed to be Midgley's biggest success of the year, even more than the fact that he actually made cyclohexane airplane fuel. The war ended before Midgley's fuel could be produced and shipped to American pilots in Europe. But the fuel research did have a practical benefit: It gave Midgley a crash course in chemistry.

After the war, Kettering asked Midgley to turn his full attention to improving the internal combustion engine. Based on the clouds of smoke and dust it emitted, both men believed it could be refined to be more efficient.

The internal combustion engine was a feat of engineering that was mostly hidden from the public because, aside from hearing the term, people rarely had a chance to see how it worked. To the average person, an internal combustion engine was simply a noisy contraption that sounded like it was loudly popping popcorn. But the technology behind it was a fine ballet of cylinders, pistons, and rods that expanded, twisted, and turned in synchronized motion to propel an automobile forward.

The engine operated like a person holding a firecracker in a closed fist. The fist would be still, but when the firecracker went off, the explosion would throw the person's elbow backward with tremendous force. If the elbow was positioned next to an open door, for instance, the explosive energy of the firecracker could be transferred into lateral energy as a door slamming. Repeat the process with several more fists,

# HOW AN INTERNAL COMBUSTION ENGINE WORKS

**1**
Piston lowers, pulling air and fuel into the cylinder.

**2**
Piston rises, compressing the fuel and air for easier ignition.

**3**
Spark plug ignites the compressed fuel. The explosion forces the piston downward.

**4**
Piston rises, pushing burned fuel and exhaust out of the cylinder.

**5**
The spinning crankshaft connects to the car's transmission, which powers the car's wheels.

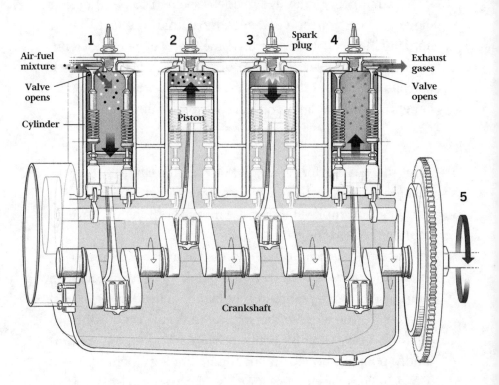

more firecrackers, and more doors, and eventually the cacophony of doors slamming, if harnessed together, could yield considerable energy. Lay the doors flat and hook them up to a rod, and the lateral motion of each slam could be transferred into rotational motion—exactly what you'd need to rotate wheels.

In a similar way, the combustion engine had a series of airtight cylinders with vertical pistons that started halfway down the cylinder and extended downward out the bottom. At their lower end, the pistons were attached to a rod, so that when the piston was forced down by an explosion in the cylinder, it would turn the rod. The rotation of the rod would force the piston back up, while another explosion would keep the rod spinning. This was the basic motion. But there was the more complicated matter of making the rod spin at different speeds, thus giving a car variable velocity. This could be manipulated by fuel quantity, which could be controlled with a pedal. For more fuel, press the pedal down. For less, let it up.

Not all fuel was the same. Gasoline in 1900 was measured by how much pressure it could withstand before igniting. In the early days of American automobiles, gasoline could withstand very little compression in the cylinders before the pressure on its molecules would force it to combust. Combustion was exactly the point of an internal combustion engine. But too much combustion, or combustion that occurred when the piston was already down, squandered the energy of the fuel without providing the automobile any additional propulsion.

Ill-timed combustion was not only wasteful; it was harmful to an automobile's inner parts. Fuel that wasn't sufficiently burned remained in the combustion chamber, where it would build up a little at a time and leave an oily residue on the pistons and cylinder heads. Too much residue would reduce the operation of the engine until eventually, like a clogged heart, there would be so much oily buildup that the engine would stop working. It was easy to tell when this was happening, because a driver could usually hear the engine clanging and clunking. These were the circumstances that caused an automobile to be dubbed an "old banger."

Early automobiles could survive this problem. For many years the excitement and novelty of an automobile was enough to paper over small inefficiencies. No one expected a technology barely twenty years old to work flawlessly.

But as time went on, the novelty of a motorized vehicle was replaced by the annoyance that it didn't always run as it should. After the Great War ended in 1918, consumers began to demand higher speeds and more powerful engines. Pushing cars to their limit—often by pushing the pedal to the floor—would result in a loud clanging sound, especially when going up hills or trying to get a quick start. In airplanes, reducing wasted fuel could be the difference between a plane flying between towns and a plane flying over an ocean, which no pilot had dared to try.

This made the problem of inefficient fuel detonation an urgent one to solve. A brainy engineer gave the phenomenon a name. He worked as a manufacturer of spark plugs, the part of the engine system that delivered the fuel, and noticed that incomplete fuel combustion sounded a lot like the engine was being tapped repeatedly with a hammer. "Engine tap" wasn't quite right, so he called it "engine knock."

Once it had a name, engine knock became a popular problem to study, with an implicit reward for whoever figured it out. A British engineer named Harry Ricardo devised a series of experiments that questioned whether the size of the compression chamber would affect the combustion that caused knocking and whether a smaller or larger chamber could eliminate the knock. After several experiments, his answer was no: Engine size was irrelevant. So his next step was to question the fuel itself. Was there such a thing as high-quality gasoline and low-quality gasoline? He did more experiments and discovered that different batches of gasoline combusted in different ways and at slightly different temperatures. With this knowledge, he created a scale. He called the gasoline's likelihood of combusting its "octane"—"oct" because methane-based hydrocarbons have eight carbon atoms. The higher the octane in a batch of gasoline, the more it could withstand knock and, by extension, the hotter and more powerfully it could combust.

The year was now 1919, and almost overnight the holy grail of automotive research became how to raise the octane of gasoline. On the octane "scale" coined in the 1920s, the era's fuel likely had an octane of around 40. It would have to reach at least 60 to eliminate knocking, and performance would increase as it rose. (Today's octane ratings range from 85 to 94.)

Someone would eventually solve this puzzle, and Thomas Midgley wanted to be first.

After the war, Midgley got to work trying to crack one of the most lucrative questions in American innovation. Kettering, who had seen how quickly Midgley could identify a problem, concoct an experiment, and drum up a solution, knew that Midgley was the man to get to the bottom of engine knock.

And that when he did, he would make them both a lot of money.

# 6

## Dayton, Ohio — 1919

Fixing engine knock was a different kind of challenge for Midgley. Almost all his formal education had been in mechanical engineering in the style of Thomas Edison, who built physical contraptions out of parts. He was most comfortable when he was formulating new ideas, building prototypes, and tinkering until a product was ready to share with the public.

Engine knock, however, needed no new contraption. Incomplete fuel combustion was a problem not of physics but of chemistry, and this required Midgley to learn as much as he could about how chemicals reacted together and changed states. Copying German jet fuel had been easy in comparison. This next challenge would drop him in the open ocean of science.

Chemistry was the youngest of the natural sciences. Unlike biology and physics, which had been around since the ancient Greeks, the building blocks of matter had barely been laid a century before Midgley was born.

Generations of scientists had debated whether chemicals, such as gold, iron, and copper, had an identical composition everywhere on earth. When they concluded that the metals did, it meant that all instances of those chemicals must have the same so-called atomic weight. The next step was to measure the atomic weights of chemicals (or elements), which yielded another striking finding: A relationship existed between an element's weight and its behavior.

To put it another way, when the known elements, roughly thirty in all, were lined up, the first element had similar properties to the eighth element, and the second resembled the ninth. Arranging them into columns effectively created a treasure map. Between the known elements there were blank spaces, which predicted exactly which elements (including their precise weights) were still waiting to be found. The table became the foundational map of chemistry, and the way it was arranged led to a simple diagram that would one day line the walls of science classes everywhere.

Elements were easy to predict, but they were difficult to actually find. In 1900, there were eighty-three known elements. Over the next twenty years, scientists added just three more.

Molecules, though, were far more common. Molecules are collections of multiple atoms, like how water is made up of the elements hydrogen and oxygen ($H_2O$) or how methane connects one carbon atom to four hydrogen atoms ($CH_4$). These are small molecules; atoms sometimes bond together into massive molecules with hundreds of atoms. One of these large molecules would hold the solution to engine knock.

Midgley's problem—or, rather, his first problem—was that he was not a chemist. He had taken just one course in chemistry in high school and another one at Cornell. It never occurred to Midgley while studying that he would need to know the first or last thing about how molecules arrange themselves in a drop of water, let alone a gallon of fuel.

Except something had changed in the years after he graduated from the university. It wasn't that all the good inventions had already been invented, as the U.S. patent commissioner Charles Holland Duell was long credited with having said in 1899. Rather, the scale of invention had begun to shrink. Innovation was shifting from creating full-scale objects to trying to make their small parts and pieces incrementally better, like improved insulation to make wires more conductive, or better oils to make an engine run smoother. While the world was hypnotized by big-ticket inventions like air-conditioning, windshield wipers, and electric blankets, a series of more introverted inventors were working quietly in their laboratories, making unflashy things like rechargeable nickel-zinc batteries and a toggle for electrical currents in the home to turn on the lights. (As it happened, some of the biggest inventions still to come were indeed tiny, including penicillin in 1928, pregnancy contraceptives in 1938, and polythene, otherwise known as plastic, in 1933.)

The day Midgley realized this, he knew he needed to think like a chemist. He carried a copy of the periodic table in his breast pocket, and in the evenings he practiced balancing chemical formulas to teach himself how molecules transformed.

A major pillar in chemistry, just as in physics, is the conservation of mass and energy. In every reaction, regardless of its complexity, every material is either maintained or transformed. Suddenly it was easy for Midgley to see chemistry in almost every mundane activity of life. Fruits ripened in his family's fruit bowl because glucose and oxygen converted into carbon dioxide and water. Green leaves turned red in the fall when chlorophyll drained away and sugars converted proteins into new pigments. Metal rusted because iron combined with oxygen, and lighting a match caused a flame because striking it caused red phosphorus to turn to white phosphorus, which has a lower ignition temperature.

Knowing these basic properties, he sought to map out how a car engine worked. Internal combustion was simple. Under ideal conditions, an engine combined fuel and oxygen and converted them into carbon dioxide and water, releasing enormous energy in the process.

But no chemical reaction ever happened under ideal conditions. Midgley knew that combustion in practice was far more complex. Innovating on his earlier combustion camera, Midgley built his own combustion chamber with a small piece of glass on one of the cylinders so he could peer into the engine and watch engine knock up close. And once he could, he could swap in different chemicals to find the perfect solution.

Even before Midgley started experimenting, there was already a substance on earth known to dramatically reduce engine knock. Ethyl alcohol, the same substance as in the beverage, was a powerful fuel since the early nineteenth century, when it was combined with turpentine to power lamps. Henry Ford used ethyl alcohol in his early tinkerings with cars. So did the American and German developers of the internal combustion engine Samuel Morey and Nikolaus Otto.

Ethyl alcohol, later shortened to "ethanol," was simple to produce from farm products like corn and sugarcane. It worked so well that every inventive mind of the era— including Ford, Alexander Graham Bell, and even Midgley and Kettering—believed that ethyl alcohol would be the "fuel of the future."

Midgley was so excited about ethanol that he would later drive a car powered by a gasoline-alcohol blend one hundred miles from Dayton to Indianapolis to a meeting of the

Society of Automotive Engineers, where he declared to his friends and colleagues, "Alcohol has tremendous advantages and minor disadvantages."

One of those disadvantages, however, was that there wasn't enough of it. In 1919, America had enough sugar and corn to offset the country's gasoline consumption for two years, but that would mean using *all* of it, which seemed a poor use of American land that would be better spent growing food for people to eat. The next best option was to make ethyl alcohol from cellulose, a chain of glucose monomers in thrown-away items like sawdust, newspaper, seaweed, and cornstalks. There was plenty of *that* kind of waste, but the process to transform it into burnable alcohol required substantial energy—and, thus, substantial money. Going that route would boost a gallon of gasoline from twenty-five cents to a stratospheric two dollars.

However, it was reasonable to think that America's technology would one day catch up and bring down the price of ethanol derived from cellulose. Or, better yet, that time would yield an entirely new fuel that didn't rely on ethanol *or* gasoline. Which was for the best, because, from Henry Ford on down, almost everyone involved in making and selling automobiles in 1919 believed that the United States had twenty years of oil remaining, thirty at most. The U.S. Geological Survey estimated that America had only 7 billion barrels of oil still in the ground, and with American consumption steeply rising, it wasn't hard to do the math. (As is now clear, the agency's assessment was wildly off: since 1922, the United States has produced more than 240 billion barrels of oil.)

Mindful of the twenty-year deadline, Midgley convinced Kettering that whatever anti-knock fuel they could find would simply be a "bridge fuel," as he called it, to increase the power and longevity of a gallon of gasoline until a brighter future arrived when cars would be powered by a better fuel that was limitless and cheap. The two men even discussed whether the magic fuel of the future would be the sun. "Energy expended by the sun on one square mile of the Sahara desert would operate all the motors of the world," Kettering once speculated.

Kettering found Midgley's bridge argument to be sound, at least in theory. He didn't have every detail worked out, but that didn't stop him from declaring his thoughts in his usual distinct, strange way of speaking. One day in 1920, at a luncheon with his engineering team, the men ate bean soup. Kettering peered down at his soup and then broke into the conversation.

"You know, fellows, a bean is pretty smart," he said to the quizzical looks of the table. "Nature provides the bean with a quantity of nourishment to keep it going until it gets a start in life. When planted in the ground, it sends up a sprout to take a look around. There it could say, 'I'll just grow in this lovely sunshine and put out a lot of leaves. I have plenty of bean meat to keep me going for a while.' But the bean, being smart, does no such thing. Instead, it uses its store of nourishment to send roots deep into the earth. Only then it is ready to put out leaves in the sunshine."

The anti-knock fuel that Midgley was working on was, in Kettering's words, "nothing but 'bean meat' to keep us going until we can get a good start."

He drove home his bizarre point as the luncheon concluded. "If we are as smart as the bean, we will, while petroleum does last, dig into the secrets of nature. If we do that, we will find other sources of energy to keep us going after petroleum has been used up."

Back in his lab, the first substance Midgley tried worked. Not well, but it worked.

Midgley wanted to dye gasoline a darker color to get a better look at the normally light-colored liquid in the combustion chamber. He went to the stockroom and asked the attendant for some oil-soluble dye. The man didn't have any. He only had water-soluble dyes, which would be of no use in gasoline.

"Try this," the attendant said to Midgley, and handed him a bottle of iodine.

Iodine (I) was nothing more than a dye that turned the light-green gasoline brownish black. When Midgley poured it into the engine, the knock disappeared. Iodine was all it took.

Except iodine wasn't the answer. For one thing, a tiny quantity cost four times the price of a gallon of gasoline. If gasoline was twenty-five cents, adding a dollar was out of the question, even if it eliminated knock. Iodine also reacted with metals in some car parts, including the carburetor and gasoline piping, and made them disintegrate.

But that first experiment proved that an anti-knock compound did exist, which convinced Midgley that, in the hundreds of other bottles in the stockroom, surely some would fix the issue.

None of them did. Midgley had two assistants, Thomas Boyd and Carroll Hochwalt. Within two weeks, they had mixed a few drops of every other sample in the stockroom into gasoline before pouring it into the test engine. None made a lick of difference to the engine knock.

After a month, Midgley began writing to industrial companies that made more elaborate chemicals, ones not found in a hardware store.

Midgley tried aniline ($C_6H_5NH_2$) the same day it arrived in the mail. It worked—much better than iodine, in fact, and it was much cheaper. But it, too, had an "appetite for engine parts," as Midgley put it. And it smelled terrible.

"I doubt if humanity, even to doubling of fuel economy, will put up with this smell," Midgley told Boyd.

Then he turned to bromine (Br), carbon tetrachloride ($CCl_4$), nitric acid ($HNO_3$), and hydrochloric acid (HCl). All of those seemed to *increase* knocking when added to the fuel and air mixture.

More weeks passed without any breakthrough. Midgley grew frustrated. Desperate for more compounds to try, he asked Kettering to require that if any new chemical came into the office for any reason at all, it must first be brought to him to try as an anti-knock compound. Via this route, he obtained a small glass bottle of selenium oxychloride ($SeOCl_2$), a solvent for glues and gums. It reduced knock even better than aniline, but it, too, turned engine parts into chemical sludge.

Since selenium seemed to be a worthwhile element, one of the men had the idea to substitute selenium oxychloride with diethyl selenide ($C_4H_{10}Se$), two substances so distinct from each other that the substitution was like replacing a blue sweater with blue shoelaces. This reduced knock better but had a major problem: Selenium smelled like rotten horseradish. Midgley tried to replace the selenium with tellurium (Te), which seemed to work even better—until he got home that night.

"Where have you *been*?" Carrie Midgley asked, wrinkling her face.

"Out to the country club," he said.

"What did you *do*?"

"Nothing out of the ordinary."

"What did you *eat*?"

"Nothing special."

"What did you *drink*?"

"Coffee and water."

Carrie made her husband sleep in the living room. The next morning at breakfast, Midgley's kids refused to come near him.

Back at the lab, Midgley's two assistants told him they had also been put off at home. One of Midgley's colleagues told him that his family had taken to "greeting me with gas masks on." None of the men could understand it. They didn't think they smelled.

It was the tellurium, they realized, which had a "satanified garlic odor," one of the men said when he sniffed it in the bottle. Another joked that it smelled so bad that the scent of a skunk would be refreshing in comparison. During their experiments, the tellurium had vaporized and become embedded in their skin and clothes so subtly that they didn't even notice. But even slower was the process to get it out. For two weeks, no quantity of baths or laundering could eliminate the devilish odor. Three weeks passed before Carrie allowed her husband back in their bedroom.

Midgley spent the next year carrying around the scents of other industrial chemicals. Some smelled like oil, others like sulfur, and some like garlic. As more time went by, Midgley thought to substitute tellurium with tin, since they were in the same row on the periodic table and nearly the same atomic weight. To liquefy it, he thought to pair tin with the alcohol solvent ethyl in its four-molecule arrangement, thus making tetraethyl tin ($C_8H_{20}Sn$). This worked as an antiknock, but not completely, and when he went to price it out, tin, like dozens of other chemicals he had already tried over the prior few years, was too expensive.

Midgley and Boyd could tell they were close. The search had narrowed, and looking at the periodic table gave clues as to where the investigation would end. It had become "a period function," in the words of one historian. Or, in Midgley's view, "a fox hunt."

The pair proceeded with frenzied excitement, like treasure hunters who had uncovered the final clue and now had to dig beneath the marked X. The discovery that tetraethyl tin was a reasonable anti-knock compound suggested that the answer could be found *around* tin. They had already tried several other metals known on the periodic table as "post-transition metals," with low vaporization points, including gallium and indium. But they hadn't looked *below* tin.

And right below tin was lead.

If tetraethyl tin didn't work, could there be something to tetraethyl *lead*? Tetraethyl lead—$Pb(C_2H_5)_4$—sat comfortably on the list of known but not terribly exciting compounds. It had been around since the 1850s, when the great chemists of the world were locked in a race to develop a class of chemicals that could make a carbon atom bond directly with a metal, otherwise known as "organometallic compounds." These compounds would be powerful catalysts to speed up chemical reactions in factories and manufacturing plants. No one knew at the time that organometallic compounds would help spark the industrial revolution.

In the 1850s, the discovery of a lead atom conjoined to four ethyl groups—$Pb + (CH_3CH_2)_4$—had no notable qualities. In fact, not long after a German chemist named Carl Jacob Löwig published his discovery of tetraethyl lead in 1853, it faded into chemical obscurity, buried on the backlists of chemistry journals and periodicals with the names of other chemicals that seemed to have no discernible use.

Midgley and a colleague spent a day making tetraethyl lead. The ethyl was important. Without it, the lead would not dissolve evenly into the gasoline. When the lead and ethyl came together, it formed a cloudy white liquid, like wet glue.

And then . . .

On December 9, 1921, they fed the new fuel into the test engine—in a ratio of 1 percent tetraethyl lead to 99 percent gasoline. It was a much lower quantity than any other chemical they had tried. They figured they could always add more.

They stood back and watched.

Mostly, they listened. The sound of the engine grew deeper, like a motor that had found its rhythm. Something remarkable seemed to be happening.

Midgley's eyes got wide. Another moment passed. He let out an excited yell.

Midgley darted from the room to find Kettering, who was already on his way to the floor after hearing the men yell. The two ran into each other in the hallway.

"Boss!" Midgley said. "Listen! Hurry! Come quick!"

They rushed back to the engine, still purring deep and low.

The celebration lasted several days. Midgley, Boyd, and Hochwalt "danced a very unscientific jig," Midgley later recalled. Midgley had been working on engine knock for more than three years. He had tested 143 chemicals, or 15,000, or possibly 33,000; the numbers seemed to grow with every retelling. His experiments had taken him deep in the periodic table to compounds he knew intimately, and others he barely understood. He had used the Edisonian method of "cut and try," a.k.a. trial and error, until he reached the solution.

From a distance, the boy who wanted to grow up to be Thomas Edison had notched a considerable breakthrough. He had proven to his bosses and his colleagues that he was a certified inventor, possibly one of the brightest innovative minds of his era. His discovery would indeed change automobiles, maybe even the world. But how, he could not yet fathom.

Word of Midgley's discovery got around. Within days, anyone in Dayton with any knowledge of automobiles had heard about Midgley. Within weeks, word spread to Detroit and New York City. Every car company was looking for innovations that could set them apart from the pack, like new wheels, more durable tires, and engines with eight cylinders, even ten. But anti-knock fuel was a breakthrough every company wanted. It would mean their cars could run cleaner and go farther and, in the process, boost the allure of the automobile.

In Detroit, the bosses at General Motors were especially enthused about the news from Dayton. GM was a big company, but it was still small compared to the Ford Motor Company. In December 1921, the same month that Midgley first tested tetraethyl lead, GM reported selling 193,000 cars—a measly quarter of Ford's 845,000.

When news of Midgley's tetraethyl lead discovery made it up to the desk of General Motors CEO Sloan, the boss was initially unimpressed. Kettering had been telling Sloan for months that Midgley had found incrementally better anti-knock compounds. Sloan likely thought that tetraethyl lead would also be quickly replaced by something modestly better a few days or weeks later. A lawyer for General Motors later recalled Kettering's routine hype and Sloan's skepticism: "When Kettering found that the element iodine would do it, he said, this is the answer. And when he had aniline, he said, this is the answer. And when he had selenium, he said this is the answer... And so, when tetraethyl lead was discovered, Sloan thought: 'it won't be long before we get something better than this.'"

But Sloan's skepticism turned to intrigue when he learned about Midgley's follow-up experiments that proved that tetraethyl lead was even more favorable than it first appeared. Wanting to find the minimum amount of tetraethyl lead required to stop knocking, Midgley weakened the amount of the substance until it appeared that he only needed *one-fortieth of 1 percent*—or barely two or three grams in one gallon of gasoline. It was a minuscule quantity of a miraculous substance that even the curmudgeon Sloan began to see as a game changer.

Amid all the celebration, it was months before anyone thought to ask any questions about tetraethyl lead. As far as Midgley, Kettering, and Sloan knew, it was simply an old compound turned useful in a new era, no different from a vintage wine plucked from an overfilled cellar. Questions about what the substance actually was or what effect it would have on millions of cars and billions of people could wait.

By early 1922, Kettering and Sloan could barely conceal their enthusiasm. Kettering had a vision, and the horizon looked bright. His self-starter had already proven that transformative ideas could be unspeakably profitable. It took Kettering less than a week to do the mental calculation that charging a small royalty for tetraethyl lead—maybe a nickel on every new gallon of gas—would quickly add up to millions of dollars.

The name would have to change, though—not because lead was an unpopular word but because tetraethyl lead was a clunker that sounded as boring as the industrial chemical it was. It needed a refreshing brand name that could not only be remembered but could also be trademarked.

There in his office, Kettering held the world's first test tube of tetraethyl leaded gasoline.

"We've got to have a name for this stuff," Kettering told the room. "What do you think of calling it Ethyl?"

Two months later, in February 1922, Ethyl was announced to the world as the definitive answer to the engine knock troubles of drivers everywhere. Midgley and Kettering were excited, but both men's enthusiasm was a small spark compared to the burning joy of Sloan and the PR division of General Motors, which had a product in its hands so great that it warranted endless embellishment. Within days, General Motors' attorneys applied for a patent.

On February 24, 1922, the Dayton lab of General Motors sent out a news release about Midgley's discovery. Two days later, the Associated Press published a bulletin that reported that there was a new gasoline compound that not only would cure engine knock but—based on nothing but the hype of corporate marketers and the hope of eager consumers— would double the gas mileage of every car that used it.

This was brilliant news. A person could see no downside in betting on the future of America's automobile industry, and especially on one of its biggest public companies.

By the end of the year, General Motors' stock was up 68 percent.

7

*Alice Hamilton in 1924*

## Cambridge, Massachusetts — 1922

The faculty apartments at Harvard were the natural place for Hamilton to live in Boston. But when she visited, she found them to be a dirty and loud bachelor house for unmarried male professors. If she lived there, she would be "very finicky," in her words, about cleanliness and interruptions, not to mention the unsubtle expectation that she tidy the house and help with the cooking.

Instead, she found a room in the home of a Boston surgeon named Ernest Codman and his wife, Katherine, at 227 Beacon Street. She had two rooms on her own floor with her own bathroom and a tin bathtub set into black walnut paneling. It was much nicer than her room at Hull House, and it reminded her of her grandfather's house where she grew up in Fort Wayne. The Codmans were delighted to have her and enjoyed the way she could discuss medical matters with Ernest and more feminine interests like fashion and theater with Katherine. They invited her to join them every night for dinner.

But once the pomp, circumstance, and controversy of her appointment as Harvard's first female professor faded, Hamilton found her faculty position lonely. Her idea to only teach one semester per year, which would allow her to maintain her residency at Hull House the other half of the year, made her unreliable for large projects. She could not lead a department or edit a medical journal, stunting two common paths to seniority.

In some ways, this was disappointing. Not starting major pursuits at Harvard guaranteed that she would not finish them, which gave her a foreboding sense of failure. She later wrote in a letter to a friend that she felt she was falling short.

"I cannot think of one thing that has gone well and I know that the fault is nobody's but mine, that I came brashly into a milieu to which I was not adequate and tried to fill a place which needed more brains than I have. There is no use lamenting it or playing that it is not true."

But in other moments, Hamilton was grateful not to be too tied to academic life. Harvard had a pecking order that caused men with extraordinary minds to do tedious things, like wait their turn and pay their dues. Hamilton was forty-nine when she got to Harvard, and at that point she had little interest in attending receptions and making mindless chitchat with junior professors twenty years younger.

Besides, Hamilton had settled into a nice rhythm shuttling between Boston and Chicago. She was busy keeping up research, writing, and friendships in both places to such a degree that letters to her addressed to one location would occasionally arrive after she had already left.

She seemed to fill every spare moment with a constant barrage of conferences and lectures that requested her attendance. The extra money helped, but it often came with special caveats that would not be required of men. For every $25 she earned from a lecture at Connecticut College or the University of Pennsylvania, she had to undergo two or three days of social engagements, lunches with town ladies, evening receptions, and endless meetings. "You almost perish [of boredom]," she complained to her sister Margaret. She hated small talk and was frustrated by the social expectations asked of a woman. But if she politely declined such invitations, the requests turned into begging. Surely you can't deprive our wives of such stimulating discussions, she was told.

By 1922, her well-managed calendar had grown into a nest of hurried commitments: Wilmington for two days followed by New York for five, then Bryn Mawr for two, and back to Philadelphia. In Boston for a week, she was greeted by a letter from an editor at Macmillan in New York who wondered whether Hamilton would like to write a book about her work. As with all the other offers, Hamilton committed quickly. She appeared to feel a strong sense of obligation, and that the slightest slipup could threaten her position at Harvard. She described the freelance opportunities to her sisters and friends with a common refrain: "I felt I could not refuse."

In March she made two visits to Chicago, and in the spring she made more than a dozen trips around the Northeast to New York, Providence, Bar Harbor, Washington, and Hadley, Massachusetts. Idling away the hours on trains, she'd jot notes of stories and case studies to include in the book, and when she got to Chicago, she'd unfurl them all in her "work-shop," a nook she reserved for the summer in the back stacks of the John Crerar Library on the corner of Randolph Street and Michigan Avenue.

Finally, in July, she retreated to her country house in Had-lyme, Connecticut. She enjoyed the intense peacock blue of the flowing river, and she would take long nature walks on a small island in the river where brown cows grazed beside yellow wildflowers. She took watercolor lessons and found painting to calm her mind.

Hamilton was now fifty-three, an age that both surprised and exhausted her. So much about her life was unconventional, and living such a contrary existence required frequently fending off rude questions and suggestions. To the outside world, she was Dr. Alice Hamilton, esteemed doctor, professor, and scientist. But more intimate parts of the world also saw her as pitiful Alice Hamilton, a woman who violated her gender's conventions in two bemusing ways: having a job and not having a husband. In her twenties and thirties, she was constantly asked when she might settle or what kind of husband she was looking for. Her polite refusal of such advances, even from well-meaning friends, left many with the impression that there was something peculiar about her, something that should not be named but in the politest sense could be described as a regrettable condition of not being interested in men.

Hamilton was indeed not interested in men. Since she was a teenager, Hamilton and her sisters professed deep apathy about male counterparts. One reason was that Alice was more drawn to intellectual than physical stimulation. She preferred friends with whom she could speak freely about religion, books, and current events. According to one biographer, she felt that "such friends could rarely be found among the opposite sex." The other reason was that she had seen little evidence that men were worthy of great attention or commitment. She recalled the embarrassment she felt as a child when her own father, Montgomery Hamilton, drank too much and had to be regularly helped home by neighbors. After he frittered away large portions of the family's money on alcohol and bad investments, Hamilton and her sisters concluded that their father demanded more attention than they wanted to give. Over time, Hamilton seemed to extend this view to all men.

On the rare occasion that she agreed to a date with a gentleman—usually after a painstaking ordeal of a setup that she could not refuse—she returned home and wrote a screed to one of her sisters about how painfully boring the evening had been. In one of the worst, she recalled dining with a doctor who, in the course of one dinner, revealed himself to be "cruel, neglectful, lying, and utterly callous to anything but the interests of the company he was working for."

Part of her disinterest in men also seemed to come from a more general sense that a husband, even a nice one, would stunt her career and fill any spare time with domestic drudgery like ironing collars and boiling eggs. "There seems to be no valid reason to refuse anything when . . . the day seems so empty without a lot of engagements," she wrote to one of her sisters. She sensed that if she stopped kicking underwater, she would lose her momentum, and in short order new triumphs would pass her by.

Hamilton was good at keeping in touch. She often had several dozen pen pals at a time. Her most frequent correspondents were her three sisters—Edith, Margaret, and Norah—and her two cousins, Agnes and Jessie. They all grew up together in the family mansion in Fort Wayne and maintained a life-long sorority of sorts, confiding in each other about their nonconformist lives, both professional and personal. The sisters all worked with children: Norah as a teacher, Margaret as a school headmistress, and Edith as a former teacher who was transitioning her career to write books about the ancient Greeks. Her cousins pursued different paths, Agnes as a social worker and Jessie an artist. But their similarities were enough to bind them closely together. Their greatest commonality was perhaps that none had a male partner. Margaret had a female one, named Clara Landsberg, who was inducted into the group and became, in the words of Alice, "one of us."

Hamilton also had remarkable connections to powerful people. Whenever she passed through Washington, D.C., she had a standing invitation to join the House of Truth, a cooperative on 19th Street near Dupont Circle that started as a group house for young men in government service and, as they rose in seniority, turned into a political salon. Almost every night, boisterous discussion filled the house. Many of the debates were about the problems of industrialization and how the government could improve the conditions of laborers, sweatshops, and immigrants. The intellectual intensity of the roundtables—challenge and be challenged—strengthened the ideas of people eager to rise to higher office.

"The talk at the table must always be general, tête-à-tête conversation was frowned upon," Hamilton recalled of the great fun she had discussing substantive matters with the men in a way she felt she could not with her Harvard colleagues. "This made for a very lively conversation which could never degenerate into dullness because one of the hosts could always catch the ball and throw it to a good catcher." It was here that Hamilton sparked friendships with a young law professor named Felix Frankfurter, who would later become a Supreme Court justice, and an ambitious journalist named Walter Lippmann, who would rise to the top levels of American media.

But Hamilton was not snooty in her relations or hungry for power. She kept in touch with any willing correspondent, even those with whom she had little in common.

One of her most curious correspondents was a man in St. Louis named Frank Hammar, who had written to her so many times that Hamilton now considered him a friend. A decade prior, while Hamilton was conducting her shoe-leather epidemiology in Chicago, Hammar, who owned the Hammar Bros. White Lead Company in Missouri, sent her a letter. He was having the same troubles in his St. Louis factory that Hamilton was investigating in Chicago smelters. Too many of his men were either quitting or dying. A doctor at the company suggested that Hammar make the workers wear rubber gloves. He did, but the reforms that Hamilton suggested in her final 1911 report—uniforms, masks, ample ventilation— made a much bigger difference. Hammar was so impressed with Hamilton's ideas that he kept sending her letters even after he put the precautions in place.

Hamilton thought Hammar was a good egg, especially in an industry filled with neglect and complacency. He was the rare company boss who seemed to care about his workers and the effects of known workplace toxins. Their blooming friendship proved helpful ten years later when Hamilton got an opportunity to finally grow in her role at Harvard.

Two professors visited Hamilton in her office one day in late 1921. Dr. Joseph Aub and Dr. Cecil Drinker were young pathologists in the medical school, both in their early thirties and eager to make their names in the emerging field of public health. They explained that they had come up with an idea and needed Hamilton's help.

What they really needed was money to fund a series of research experiments. They lavished praise on Hamilton for her "socially conscious" research of lead in Chicago, by which they meant the way Hamilton collected anecdotes to prove workers were poisoned by their jobs. Aub and Drinker said they wanted to go a step further and study lead from a "scientific standpoint," by which they meant the physiological impact of lead on the body. Humans had known lead was a poison for thousands of years. But Aub and Drinker wanted to find, for the first time, the exact quantity of exposure that made a person get sick and then how long that person had to maintain continued exposure before they'd die.

This was a potent question in the early 1920s. Lead was an obvious danger when consumed in large quantities. But two things began happening in the twentieth century that raised new questions about the use and impact of smaller quantities of lead. For one, lead was popular again. Between 1900 and 1920, lead mining in the United States tripled. It was central to the booming automobile industry, used to make wheels, axles, and batteries, and once large drums of it were sitting in factories, it was a convenient and accessible material to make other vehicle devices, like wheel weights and fuel lines. And secondly, rapid urbanization and the growth of America's military required more lead for plumbing and ammunition, and industrial factories began to churn out millions of dollars in pottery, cooking utensils, and cosmetics.

As lead became embedded in American life, more reports of lead poisoning began to surface. Parents grew accustomed to children being sent home from school with terrible stomach pains, or mothers had strange redness where they had applied makeup, or fathers complained of fatigue and constipation before it progressed into pain in their kidneys and frequent spells of confusion. The telltale sign was a bluish line on a person's gums, a mark later given the name "lead line." A ballooning number of cases of moderate lead poisoning demonstrated that, in addition to being harmful in large quantities, lead could be damaging in small quantities too.

But how small? Even the top scientists at the top medical research university didn't know. But they wanted to find out.

Aub and Drinker explained to Hamilton that they needed help designing, conducting, and especially funding an in-depth investigation. They also needed a source of lead that they could use in experiments, in which they planned to give lead to different animals, like dogs and cats, to see how much accumulated in their muscle tissue. Most importantly, they needed access to lead-smelting facilities where they could take real-time measurements.

Hamilton offered to write to her friend Hammar. Not only was Hammar sympathetic to such investigations, he had prac-tically begged Hamilton for one. He said he was "genuinely concerned" about whether his business was poisoning his workers. Even more helpful, Hammar also had close ties to the Lead Institute, a trade association that represented the four major American lead-producing companies: the American Smelting and Refining Company, the National Lead Company, the Anaconda Company, and the Hecla Mining Company.

Hammar responded that he was happy to hear from Ham-ilton. And, sure enough, he made the necessary inquiries, and after he personally vouched for Hamilton as his friend who had made helpful suggestions in the past, he got the as-sociation to commit $52,500 for a three-year study of the oc-cupational hazards of lead. The money was soon on the way, he said, but there was only one catch: Since the industry was funding the study and was permitting access to several com-panies' facilities that processed lead, some executives wanted to see in advance how the experiments would be designed—as a courtesy. It went without saying that they would be able to propose changes to anything they found objectionable.

Aub and Drinker found the request reasonable. So long as their findings would be arrived at entirely via the scientific method, they felt they had "full scientific liberty."

Once they accepted the money, they thanked Hamilton profusely, giving her rare praise for an accomplishment that was not only major but also something the men could not do on their own. "I don't know how she did it," Aub later told an interviewer. "What makes her performance all the more remarkable is that before Alice Hamilton came on the scene, many leading people in the lead industry would not even admit the existence of lead poisoning."

Hamilton enjoyed the win, but the admiration from men at Harvard was especially sweet.

The "lead study," as it would become known in the hallways of Harvard Medical School, began in April 1921. After Aub and Drinker had declared their plan to research one of the most dangerous, destructive, and widely used materials found in the natural world, almost a dozen colleagues, young researchers, and even undergraduate students offered to help.

Designing the installments of the study required a wide scientific view, encompassing components of chemistry, biology, physics, and more than a dozen related disciplines, like mathematics, statistics, pathology, hematology, and physiology. This would be far more complex than simply mixing two chemicals and seeing what happens. Hamilton helped recruit several chemists, cell biologists, and pathologists to contribute to what was becoming a fast-growing research collective.

After more than three months of planning, Aub decided that the first two installments would explore how lead interacts with biological material like a person's liver, kidneys, and urine. This required finding a way to measure very small amounts of lead in samples of blood, urine, and muscle tissue. In the summer of 1921, a young chemist in Aub's lab named Lawrence Fairhall collected samples of muscle tissue from healthy rats and rabbits that had been freshly killed and applied atopic absorption spectrophotometry, a technique that used light radiation to find trace amounts of metal. Fairhall worked quickly. Even though the samples were no longer attached to a body, they were still biologically able to perform certain functions for a few hours, like breaking down drugs or toxins.

Fairhall added small pieces of lead to the dishes of blood, urine, and muscle in front of him. Then, after a short pause, he would pass a small beam of light over the sample to see how much of the light was absorbed by the lead atoms that remained. The rest, he concluded, had been metabolized by the liquid or the organs. His findings surprised him: Very little of the lead was metabolized quickly, suggesting it stuck around for longer. Fairhall had made two major discoveries: One, that tiny amounts of lead appeared to change the chemistry of every biological material. And two, that it was possible to study lead exposure in smaller increments, like breaking down minutes into milliseconds.

Months later, when Fairhall revealed his findings, Hamilton and Aub thought that the next step was to investigate how much lead it took to have adverse effects on the structure and function of red blood cells. Blood was the body's primary conveyor belt. If lead was ingested or inhaled, blood cells would reveal how—and how fast—it would be distributed to other organs like the heart or brain.

Unlike Fairhall's experiments, which had required only petri dishes of urine and blood, this next question needed live animals. One afternoon, Aub purchased two dozen rats and two dozen rabbits from a local pet store. In later experiments, Aub would expand to cats and dogs.

Starting in March 1922, Aub began to feed and inject the rats and rabbits with small grains of lead. He added lead acetate to their food and water and allowed the animals to live unbothered for several weeks. During that time, he watched how they behaved. The rodents began to move noticeably slower and sleep for longer periods. The rabbits, which formerly hopped around enthusiastically when receiving carrots and grain, seemed to lose interest in any stimuli.

After the exposure period, when the animals showed unmistakable signs of lead poisoning, Aub killed them to investigate their inner organs. In almost all the animals, large lead deposits appeared to have stopped the functioning of the kidneys and liver, the organs responsible for synthesizing cell proteins. Without proteins in cells, almost every organ and bodily system, from digestion to respiration to even basic brain function, would quickly break down.

Left unanswered, though, was how much lead it took to affect red blood cells. The informal consensus was 40 parts per million. But Aub wondered if it could be lower, so he conducted several dose-response experiments. He started at 40 parts per million in a sample of blood and incrementally lowered the exposure each round, to 30 and then 20 parts per million. Lead was still corrupting red blood cells, even at the lower levels. He dropped it to 10 and then 5 parts per million. Finally, in late 1922, he arrived at the furthest possible terminus. At 1 part per million—a quantity of lead amounting to *one-tenth* of a single drop of blood—Aub could still see changes in red blood cells. There was no acceptable limit: Lead in even the tiniest quantity appeared to have a destructive effect.

As often happens in science, such a remarkable finding brought up new questions. With conclusive evidence of how lead disrupts the body, the Harvard scientists wondered what would happen if they changed a few variables. Instead of the animals *eating* tiny pellets of lead acetate, Aub wondered what would happen if they *breathed it in*. Aub used a lead chamber that was connected to a generator that produced lead oxide dust. The animals were placed in the chamber for several hours a day to simulate occupational exposure, as though the rats and rabbits worked in a smelter.

Weeks later, when Aub killed those rats and rabbits and analyzed their tissue, he was shocked to find that the animals that had been exposed through inhalation had much higher concentrations of lead in their lungs and livers than the animals that had eaten it had in their stomachs. He replicated the experiment twice to be sure, but the outcome was the same. There was now concrete evidence that the most dangerous way to encounter lead was to breathe it.

By the time they were finished several years later, the lead study had ballooned into a twelve-part series of experiments, trials, and discoveries that involved almost fifty researchers. Once the scientists demonstrated their competence with the funding from the lead companies, money came in from the National Research Council, a new wing of the National Academies of Sciences, Engineering, and Medicine that allowed the U.S. government to harness the expertise of the scientific community.

Hamilton did not author an installment of the series. She was too busy on other projects. Besides, she hated lab work. She often wrote to her sisters that she enjoyed the part of research that let her talk to people. She liked doing interviews and hearing stories. For a woman who had spent two decades making friends and jotting notes, there was nothing more boring than chemistry.

She also had better things to do. While the Harvard men were detecting specks of lead in 1922, Hamilton was drawn into a debate over a proposed amendment to the U.S. Constitution to secure rights for women. The legislation, known as the Equal Rights Amendment, would end legal distinctions between men and women.

Unlike many prominent women at the time, Hamilton thought the amendment was a bad idea—destructive even— and that it would nullify decades of hard-fought advantages women had fought for, like limited night work, no heavy lifting, and shorter work hours. In a debate between Hamilton and the suffragist Doris Stevens in the pages of the monthly magazine the *Forum*, Hamilton argued that the ERA would cause women to lose in two ways. To be legally considered "equal" to men would mean surrendering their workplace protections that had taken decades to win. In return, they would get a legal designation of "equality" that sounded nice but would be undermined by institutional sexism in businesses and courts that no constitutional amendment could eliminate. Put another way, they'd give up their modest gains and be left with nothing at all.

"To my mind," she wrote to a woman on the other side of the debate, "if you succeed in rescinding all the laws in the country discriminating against women and do it at the expense of present and future protective laws you will have harmed a far larger number of women than you will have benefited and the harm done to them will be more disastrous." (With Hamilton's opposition, the Equal Rights Amendment failed. Between 1972 and 2020, it was revived and ratified by more than thirty states, but it has not passed the legal threshold for inclusion in the Constitution.)

While she fought overarching battles like the ERA, Hamilton also stayed engaged in the granular work of toxic materials and worker protections. In 1923, the surgeon general asked her to visit factories that made electrical equipment like generators, motors, and transformers to see if the workers were harmed by any of the heavy metals they used. Unsurprisingly, they were. Lead, arsenic, and mercury were all commonly used, and the workers who suffered the worst health effects were those who handled electrical wire that was fused with lead solder. In her report to the surgeon general, Hamilton noted that she saw the same shortcomings in protective gear and ventilation systems that she had seen fifteen years earlier in Chicago lead factories. If she was frustrated that little seemed to have changed in more than a decade, she hid it well, often under a frenzied sense of purpose. Hamilton seemed to believe that as long as she kept doing her work, change would eventually come in the form of conscientious business bosses and stronger government regulations.

All the while she remained working on her book, the world's first textbook on industrial poisons, which would fittingly be titled *Industrial Poisons in the United States*. Like a forensic investigator, she pounded the pavement undeterred by people who closed doors in her face. Now that her name had started appearing in the paper as an expert investigator of occupational hazards, companies that formerly let her in began to dispatch staff scientists to politely evade her questions. And if they did, she'd simply find another source of information. She built an informal network of undertakers, nurses, apothecaries, and priests who often "let drop valuable leads." In the book, she included every detail of her shoe-leather research, visiting more than 3,000 people in two years to see how they worked, how they lived, and how almost every industrial material—including benzene, copper, zinc, brass, arsenic, and mercury—was, in almost all instances, making people's lives worse. Even with all the work she put in, she still felt a foreboding sense of failure that the book wasn't comprehensive enough. When she received the book's proof pages, she left the box unopened for days. "I have not even opened the wretched thing, knowing so well how ashamed I shall feel when I do," she wrote to a friend.

But nothing got more of Hamilton's attention or ink than lead. Lead was a special poison, she wrote, one to which no one was immune. Her years of investigation had confirmed that lead was harmful in almost any context, in almost any quantity, and to every bodily organ.

"Lead is the chief harmful agent" in America, she wrote. And in addition to its obvious harm in large doses, even "the danger done by repeated small doses is lasting," she declared.

When her editor finally convinced her to submit the book for publication, not because it was finished but because it was almost six hundred pages, Hamilton let it go reluctantly. It was as though she had been locked in a battle with a ferocious and highly destructive enemy, and even after she had exposed the full extent of its evil, she could not let go.

By the summer of 1922, there was no one on earth more knowledgeable about the impacts of lead on the human body than Alice Hamilton. She had developed the flare of impatience of a woman who had no time for the dillydallying or corporate bamboozling that had cost so many people their lives.

Though naturally friendly and polite, Hamilton had run out of patience. She was ready to fight with the full strength of her rhetorical, organizational, and even physical strength if it became necessary. And soon it would be necessary.

# 8

## Dayton, Ohio — 1922

The excitement about the new type of gasoline began with an opinion column by retired automobile racer Barney Oldfield in the Fall River, Massachusetts, *Evening Herald*. Oldfield was a name synonymous with speed. In 1903, he was the first person to ever drive a car at sixty miles per hour, an unconscionable velocity that warranted the headline "Wow! A Mile a Minute!"

Oldfield's credentials as the king of speed won him an invitation to the road tests of tetraethyl leaded gasoline that began in Dayton in May 1922. General Motors executives knew that not only would Oldfield conduct a vigorous test of the fuel at the maximum performance of an automobile but that if all went well, Oldfield would enthusiastically endorse the fuel in his syndicated newspaper column. Which was exactly what happened several weeks later when Oldfield announced to his readers across the country that there was "a better fuel for the motorist":

*When used, a remarkable change comes over the engine.
The engine would not knock on the steepest grade. It produced a smoothness of operation and an increase in power.
The substance, newly discovered, increases power, reduced
oil contamination in the crankcase, and above all stops the
knock. If distribution were national, we could use high-
compression engines in existing form and get far better
mileage, much more power and speed, and then slowly design special high-compression engines, lighter, more powerful and cheaper than present types. The saving to the
nation would be tremendous and to the individual it is a
Utopian motoring dream.*

Oldfield's stamp of approval was cause for excitement.
Not just for the average American motorist, who could only
dream of climbing a hill without loudly announcing it to the
entire neighborhood. But the excitement spread to the auto-
makers in Detroit, who had been desperate for a fuel that
could fortify gasoline to stretch the world's supply.

Tetraethyl leaded fuel seemed to arrive at the perfect moment to quell the anxieties of the public. Not only did it stop
knocking just as well as ethyl alcohol, the widely accepted
fuel of the future; even better, all it took for improved engine
performance and fuel economy was a quantity so small, it
would hardly affect the price of gas at all—maybe by just
three cents.

Three cents, however, could add up. And that was the part
Kettering and Midgley grew to like most about Ethyl gas. In
1922, Americans bought 6 billion gallons of gasoline. Midgley believed that if Ethyl became the preferred fuel of just 20
percent of motorists, three cents would quickly shake out to
more than $35 million per year. Only a handful of companies
in the world had ever made that much money in a year, let
alone on a single product. Naturally, $35 million turned out

to be a low estimate, but even in the fuel's early days, Midgley believed that if Ethyl could elbow its way into enough market share, it would be like a boulder rolling down a hill.

Before that was possible, tetraethyl lead needed to overcome a few obstacles to transform it from a successful lab experiment into a fuel used by millions of people. For one, it was difficult to make in large batches. Second, the fuel corroded the engine more in road use than the lab tests had indicated. It slowly accumulated on the spark plugs and bored holes in the exhaust pipes. There was also the question of how to transport the tetraethyl lead to filling stations around the country: If it was exposed to sunlight during mixing or pouring, the chemical bonds would break down.

In Midgley's mind, these were solvable problems. Engine parts could be made with hardier metals to resist corrosion. Adding another radical element to engines could give the lead something to bond with instead of attaching to, and corroding, car parts. Midgley eventually came upon bromine and added just enough to the gasoline mixture to stabilize the fuel.

Transporting it could also be figured out. Midgley thought that perhaps the new gasoline could be shipped in opaque containers. He was also enchanted by two men in 1916 who promoted a pill and a powder that they claimed could turn a gallon of water into gasoline. The ideas fell apart quickly (and one of the men was charged with fraud), but Midgley wondered if the pill concept could work for adding tetraethyl lead to gasoline in a way that was cheap, stable, and safe.

Kettering and Midgley decided that these and any other remaining kinks should remain internal for the time being and that nobody outside the company needed to know the sausage-making details of normal product tinkering. General Motors patent attorneys supported this assessment, a sign that the company no longer saw tetraethyl leaded gasoline as a scientific discovery that could be improved with public scrutiny but as a lucrative invention that needed to be protected at all costs. When Midgley sent out samples to Standard Oil, Sun Oil Company, and the Navy's Bureau of Aeronautics, he detailed the ingredients in the fuel mixture but left out the quantities of each, as though trying to obscure a recipe as notoriously secretive as Coca-Cola's.

Midgley's ingenuity in inventing the fuel and his savvy in protecting it started to win him public accolades. By late 1922, the New York chapter of the American Chemical Society invited Midgley to receive the William H. Nichols Medal, a prestigious annual award for "original and significant contribution to chemistry."

The award not only recognized Midgley as a leading mind in a growing field; it also gave tetraethyl lead its first public stamp of approval from a credible group of scientists—effectively, an endorsement that could be leveraged to promote, make, and market the fuel. And endorsements from scientists would be extremely helpful once the shortcomings of tetraethyl lead came to light.

For now, the millions of readers who had seen Barney Oldfield's column in the *Evening Herald*, the *Arkansas Democrat*, or the Columbia, South Carolina, *Sunday Record* had no reason to think there was any downside to tetraethyl fuel at all. The utopian dream Oldfield prophesied of more power, better mileage, and higher speed could become reality for everyone in America as soon as they could get their hands on it.

The first indication that Midgley's miracle fuel could be problematic came that same month, in the third week of July 1922. Since the day Midgley told Kettering about tetraethyl lead, Kettering had been so enthusiastic about the fuel, and especially its commercial potential, that he could barely wait to start producing it. Kettering saw it as his personal mission to make Ethyl gasoline available to the public as quickly as possible.

General Motors was a big company, one of the biggest in the United States. But its business was making cars, not fuel. To introduce the new Ethyl gasoline to the public, Kettering needed to find a manufacturer that could produce thousands of pounds of tetraethyl lead. There were only a handful of chemical manufacturers at the time and only three that could produce such a quantity. Dow Chemical, originally founded in 1897 with only one product, bleach, had expanded to make a range of chemical products like chlorine for fire flares and bromine for medicines and tear gas. Monsanto, founded four years later, built its early business on making food additives like caffeine and artificial sugar.

Instead, Kettering chose the third major maker of chemicals, DuPont de Nemours, a company almost one hundred years older than the other two that had a wide portfolio of products, including one of the flashiest explosives used in bombs, mines, and warheads, called trinitrotoluene, or TNT. It helped that Pierre du Pont, the great-grandson of Éleuthère Irénée du Pont, who founded the DuPont company in 1802, also happened to make a generous investment in General Motors in 1915, and that five years later Pierre du Pont was no longer president of the DuPont chemical company but in fact, through an elaborate game of corporate musical chairs, had become president of General Motors. The association between the two companies made it an obvious pairing and also forged a corporate bond that would safeguard the trade secrets of both.

Kettering visited DuPont's headquarters in Wilmington, Delaware, on July 20, 1922. He met with several engineers and discussed with them the precise molecular construction and formulation of tetraethyl lead. During a tour of the company's campus, the men brought Kettering to a vacant building with high ceilings and told him that they could quickly and easily outfit the facility to begin manufacturing tetraethyl lead.

As he surveyed the building, Kettering seemed distracted. He asked questions that focused on urgency. "He thought speed was tremendously important at that time," one of the DuPont officials recalled of Kettering's visit. When the Du-Pont men insisted they could get started quickly, in just a few months, Kettering was pleased. He invited them to Dayton to see the small-scale operation they would need to replicate.

A few days later, the engineers arrived in Dayton to see Midgley's lab. With Kettering hovering behind him, Midgley gave the engineers a tour and explained the laborious process he and his colleagues had undertaken to discover tetraethyl leaded gasoline.

Then he went to demonstrate it. But when he mixed small bits of lead with the ethyl solution, something went wrong. He had either mistaken the quantities or failed to properly clean the test tubes of other chemicals. Like a mad scientist's lab in a cartoon, the tubes began to form vapor and boil over. "The demonstration got entirely out of hand and spewed all over every place," the DuPont official later recalled. The group had to leave quickly. Not long after, Midgley was performing another demonstration to a group of visitors when he misjudged the chemicals and small lead fragments blew into his eye—again. To anyone who had witnessed these events, they were evidence that tetraethyl lead could be dangerous, even in the hands of an award-winning chemist who had become one of the world's top authorities on making it. But that was only if the public knew about the mistakes, and Kettering kept both incidents under wraps.

Midgley's hands would soon be the subject of international public attention. But it was in his lungs that Midgley felt the effects of lead first. By the fall of 1922, barely ten months after his first experiments with tetraethyl lead, Midgley felt tired and short of breath. What he first dismissed as a cold seemed to progress into pneumonia, and after he missed more than a month at work, he started to decline speaking invitations from local chapters of the American Chemical Society that had invited him to talk about his exciting work.

"After about a year's work in . . . lead I find that my lungs have been affected and that it is necessary to drop all work and get a large supply of fresh air," he wrote to one group.

To his boss, Kettering, Midgley was blunter about what had caused his ailment: "I find myself to have suffered to some extent from organic lead poisoning," he told Kettering in January 1923.

Fatigue and low body temperature were Midgley's dominant symptoms. Deeper inside his body, however, lead was attacking almost all his organs. Lead was known to inhibit learning and memory in the brain and to cause stomachaches, constipation, and vomiting. Future research would find more insidious effects, like a weakening of tongue muscles that made it hard to speak, and rushes of hormones that would lead a person to either extreme mania or crushing depression.

Midgley had a mix of these conditions. But he wasn't worried. He seemed to see himself like a firefighter who worked in burning buildings: He had high exposure to lead particles, so *of course* he would be the first to get sick. Even so, he believed that his sickness was minor and that the public would have far less exposure than he did, rendering any credible concern about the substance's broader effect to be almost laughably overblown. Even in the depths of his illness in the winter of 1922–1923, Midgley reflected in his correspondence his optimism about the valuable substance that General Motors would soon introduce to the world.

In a letter to a scientist who wanted to perform a laboratory analysis on tetraethyl lead, Midgley wrote back almost cheerfully to minimize any caution. "It would not surprise me if in the course of using tetraethyl lead for a year that some of your men would experience a slight case of painter's colic," he wrote. Midgley disclosed that he had the condition himself, as did several of his colleagues.

Don't fret, he wrote. "There is nothing to worry about."

More letters than usual started arriving at Midgley's office in Dayton. The first ones came from newspaper reporters and the editors of car magazines who had enthusiastic questions. Was it really "double" the mileage, or could it be "triple"? one asked. An auto mechanic asked whether tetraethyl lead could also be useful in other parts of the engine system, such as an additive in motor oil or transmission fluid. The letters reflected the exuberant public belief that Midgley had invented not just a fuel mixture to reduce engine knock but a miraculous substance that could solve almost any problem.

Buried in the excitement about Ethyl fuel, however, were letters from experts who knew intimately about tetraethyl lead because they had studied its chemical properties and had seen how dangerous it could be. Four experts sounded these early warnings. Reid Hunt studied the work of adrenal glands at Harvard, Yandell Henderson studied respiration and toxicology at Yale, Robert E. Wilson studied chemical engineering at MIT, and Erich Krause was a medical researcher in Potsdam, Germany. All four wrote to General Motors in 1922 as soon as they heard about Midgley's experiments in Dayton. Kraus told Midgley bluntly that he was making a big mistake. Tetraethyl lead was "a creeping and malicious poison," he said, even worse than organic lead. It was so dangerous, in fact, that it had killed one of Kraus's former colleagues.

One letter referred to tetraethyl lead as a serious and devastating menace to public health, and yet another referenced Midgley's erroneous belief that the quantity of tetraethyl lead in Ethyl gasoline was too small to have any serious effects.

The foremost of the four experts was Yandell Henderson at Yale. Henderson was the nation's leader in a small field of medicine known as respiratory physiology, which scrutinized the process of breathing and the operation of the lungs, diaphragm, and intercostal muscles. In 1911, the same year Alice Hamilton was interviewing lead workers in Chicago smelters, Yale appointed Henderson to investigate the gases that mine workers were exposed to underground. During World War I, he was promoted to chief of the medical section of War Gas Investigations in the U.S. Bureau of Mines to document the military's use of chemical warfare. By 1922, Henderson was not only America's top holder of institutional knowledge about breathing; he could also rattle off thousands of toxins he knew to sicken the lungs, with lead among the worst.

Henderson was so alarmed by Midgley's discovery and his plans to market the substance nationally that he made a copy of the letter he sent to Midgley and forwarded it to the U.S. surgeon general, Hugh S. Cumming. Henderson and Cumming had been pillars in the progressive era's new approach to public health, which valued workers and consumer rights over the profit-seeking whims of unregulated companies. They had also worked together previously on the first federal regulations for air traffic safety.

A week later, Cumming wrote back to Henderson. Cumming was sympathetic to Henderson's alarm, but he was torn. Here was a discovery that had been proven in lab tests to solve a widespread annoyance for millions of drivers. It also promised to make cars more powerful and gasoline last longer. Such a finding was being received enthusiastically by the car-buying public. Cumming felt that the government couldn't reasonably disrupt such a popular success.

Nor did Cumming have the confidence to go to battle against one of the country's top corporations. Cumming was the nation's chief doctor, but his position was a political one. Two years earlier, Cumming was made surgeon general by President Woodrow Wilson. Wilson was a Democrat who was sympathetic to the concerns of workers and consumers. He believed that the federal government had a responsibility to limit the bad behavior of corporations. But now, in 1922, the presidency was held by Warren Harding, a Republican who felt largely the opposite. Harding believed that economic growth and prosperity were key to improving the lives of Americans, and much of his campaign platform was devoted to creating a business-friendly environment that would stimulate innovation, investment, and the job market. Even if Cumming was convinced by Henderson's urgent warnings regarding public health, he couldn't stick his finger in the eye of General Motors without also upsetting the president, who would surely relieve him of his position.

Almost the entire federal government seemed to take the same approach as Cumming to regulating the auto industry—or even thinking about regulating it. Few enterprises at the time embodied the spirit of American growth as much as automaking. The only reasonable action for President Harding or anyone who worked for him seemed to be to get out of the way and let companies regulate themselves through fierce competition on price, quality, and innovation. The thinking went that so long as the automotive industry was good for America, it was good for Americans.

Plus, despite the voluminous scientific research showing lead was dangerous, and the growing medical chorus that tetraethyl lead was a potent form of it, Ethyl gasoline had yet to kill a single person. So long as no consumers were getting sick or dying, hysterical warnings seemed like an unwelcome downer in a time of excitement and expansion.

Cumming, however, did take modest steps to look into the matter. At the urging of Henderson and several other scientists who wrote to express alarm, Cumming sent a letter to GM president Pierre du Pont.

Since lead poisoning is often the result of "daily intake of minute quantities," Cumming wrote, "it seems pertinent to inquire whether there might not be a decided health hazard associated with the extensive use of tetraethyl lead."

Du Pont forwarded the letter to Midgley to answer. Midgley responded to Cumming several days later, before he decamped to Florida for a sick leave of absence, stating that no experimental data had been taken on the matter of public health. Lacking any evidence, Midgley believed it was clear that "the average street will probably be so free from lead that it will be impossible to detect it or its absorption."

This became GM's official line about the substance, all predicated on the hopeful hunch of a thirty-three-year-old engineer who had never studied the human body or the diffusive properties of lead, and who himself was currently suffering from lead poisoning.

People dying was a high standard to meet for any potential toxin. But before long, Ethyl gasoline would have its body count. And soon enough, Cumming would have no choice but to call upon the most authoritative scientist on lead poisoning in America to look closely at tetraethyl lead and determine whether there was truly an unseen danger to the American public. And when she did, Alice Hamilton would find the answer to be yes—and it was even worse than Henderson had warned.

At the moment, though, Midgley's spirits were high. He spent that winter convalescing in Miami Beach, breathing fresh air that blew in from the Atlantic. He took walks in the afternoon and slept late in the morning. He had a fever for more than two months, but it didn't bother him. He was focused on the future of Ethyl gasoline, where the landscape was filled with three-cent royalties as far as the eye could see.

"My dear boss," he wrote to Kettering from his beachside hotel. "The thermometer I am using refuses to tell me I am normal, but I certainly must be."

He went on: "The way I feel about the Ethyl Gas situation is as follows . . . I think we ought to go after it as soon as we can."

# Part Three
# A VERY GREAT DEAL OF UNFORTUNATE PUBLICITY

# 9

*The first public sale of Ethyl gasoline in February 1923 in Dayton, Ohio*

## Dayton, Ohio — 1923

On February 1, 1923, motorists who stopped to fill their cars at the Refiners Oil Company filling station on Sixth and Main Streets in downtown Dayton were offered the chance to experience the future.

That was how Kettering framed it. While Midgley was still suffering from acute lead poisoning in Florida, Kettering personally orchestrated the public debut of Ethyl. He convinced his friend Willard Talbott, who owned several Dayton stations, to sell the experimental fuel and split the proceeds with him. Kettering arranged for a photographer to document an occasion he thought would be historic. And for maximum branding that could help future marketing, he had a sign printed declaring:

# ETHYL GAS

*Anti-knock Gasoline*

PRODUCT OF

**THE GENERAL MOTORS RESEARCH CORPORATION**

"Ethyl"—the name coined by Kettering—was advertised at twenty-four cents a gallon, compared to regular gasoline for twenty cents. Kettering believed that the informed customer would gladly pay an extra four cents for a more powerful fuel. But by midmorning, Talbott had yet to sell a single gallon of Ethyl. To get things moving, a GM employee started to market the fuel personally. He approached people filling their cars and asked if they knew about "the benefits they would derive" from using Ethyl fuel. Over the next three hours, Kettering and Talbott sold seventy gallons. By the end of the day, the number sold had topped three hundred, just in time for a newspaper reporter to arrive and witness a line of cars waiting to fill up.

This was still market research, Kettering thought, not a full-scale product launch. He didn't have a limitless product to launch anyway. To bring the fuel to market quickly, Kettering could only arrange for a local manufacturer in Ohio to make 160 gallons of tetraethyl lead and pack it in one-liter bottles to be mixed in gasoline tanks at each filling station. This process was the easiest way to get the substance into fuel without disrupting the existing distribution system for gasoline. Each liter of tetraethyl lead would treat three hundred gallons of gas, which meant a liter would last for dozens of cars. But it also meant that the task of mixing just enough (and not too much) tetraethyl lead into the gasoline fell to low-skilled filling station workers. These young men had little idea what they were handling, but they didn't worry. Why would a young man worry when he was working with a new high-tech product developed by award-winning scientists at one of America's most esteemed companies?

Within a year, two boys in Dayton who handled tetraethyl lead bottles died. When Kettering heard about the deaths, he blamed the boys for their lack of precaution. "They did not realize what they were working with," he later said, implicitly admitting that tetraethyl lead was a dangerous substance. He referred to the boys as "sissies."

To Kettering, this was a small speed bump on an otherwise bright road. After seeing the successful launch of Ethyl gasoline in Dayton, Kettering cooked up an even bigger test launch. Once Midgley returned from Florida, his symptoms subsided but not fully gone, the two men wrote letters to race car drivers who planned to compete in the Indianapolis five-hundred-mile race on Memorial Day and offered free supplies of engine-boosting Ethyl. It was a hard offer to turn down, especially for professional drivers seeking every possible advantage. When the top three finishers in the race were fueled by Ethyl, Midgley made sure reporters heard about it.

Within months, Ethyl billboards and magazine advertisements appeared in three states, advising drivers not to bother driving if it wasn't with Ethyl. "*Summer's comin', folks—better change to ETHYL*," ads began declaring in the spring of 1923.

To rationalize the higher cost, Midgley's assistant T. A. Boyd had the idea to dye Ethyl gasoline red so motorists felt like they were getting something different—and special. A subsequent ad campaign referred to the fuel as a "wine color" to evoke something expensive but worth it.

To keep up with the feverish growth and marketing of Ethyl, General Motors set up a new subsidiary within the company called the General Motors Chemical Company. Rationalizing that GM needed a formal infrastructure of people to supervise the production of tetraethyl lead, Kettering appointed himself president and Midgley as vice president. Meanwhile, Alfred P. Sloan, the CEO of General Motors, was given a seat on the board of directors.

To these three men and everyone around them, it seemed in the summer of 1923 that every necessary safety precaution had been taken and every marketing dollar had been spent to ensure that Ethyl was on its way to considerable growth. And for the moment, it was.

Alice Hamilton wrote to Surgeon General Cumming early in 1923 as well, in response to an inquiry Cumming sent asking her thoughts on the tetraethyl lead matter. Was it really the danger that the piles of letters on his desk suggested it was? Months later, he sent her more pointed questions, wondering what action she believed the government might take, if any.

Hamilton demurred at first. She considered herself an expert on lead broadly, and particularly on the conditions of those who worked in smelters and mines. As to the effect of lead on the general public, Hamilton at first could only speculate. And regarding tetraethyl lead, she said she had no specific knowledge. She was a woman of meticulous research and science who was reluctant to offer any public opinion not rooted in fact. Unlike several of the male researchers who wrote to Cumming with no apprehension about making grand pronouncements on complex issues, Hamilton seemed to temper her deep expertise and hold back. She didn't want to risk entering an arena where she could become a target and lose her stature. She would rather, in her words, "cling to the pleasantness" and keep all that she had earned.

Privately, however, Hamilton had no hesitation about the dangers of tetraethyl lead. She knew better than almost any living person that lead in any context was to be avoided. She could point to the voluminous evidence she had just amassed in her recent book. There was nothing she knew about tetraethyl lead to exonerate it from danger; in fact, much was still unknown, such as how much lead dust or gas was emitted in exhaust, how long it stayed in the air, and how much a person would inhale (and over how long) on the side of a street.

One reason for Hamilton's initial refusal to wade into the debate may have been fatigue. From the end of 1922 and through the summer of 1924, she kept up her feverish pace of travel, teaching, and speaking. She gave colorful speeches to the Workers' Health Bureau and the Chicago Tuberculosis Institute, describing her work in underground mines to rooms of transfixed men who had scarcely even seen a woman in a mine, let alone one who knew more about a mine's inner workings than they did. Her travels were like bicycle spokes: From Boston, she would shuttle almost every weekend to and from New York, Greenwich, Philadelphia, or Baltimore for meetings and talks. Then, in March, she moved to Chicago and spent every weekend radiating out to St. Louis, Cleveland, or Fort Wayne.

In June 1923, she accepted several speaking invitations in California and jumped at the opportunity to bask in the West Coast sunshine. For more than a month she conducted interviews with workers in mercury mines in Fresno and Salinas, and then she spent a glorious week in Santa Barbara breathing air from the Pacific. The ocean breezes were a welcome break from the constant clouds of smog from automobiles and the dust from unpaved roads that were choking every American city she passed through.

It's reasonable to wonder whether Hamilton might have more forcefully opposed the launch of Ethyl gasoline had she not been in such intense motion at the time. Even if she was averse to getting involved, she had become a credible government source. Since her successful survey of Chicago smelters in 1911, she had taken one-off projects for the Department of Labor and the War Department in Washington. In that time she grew from an academic researcher into a respected public health officer frequently consulted by senior government officials and the White House. Had Hamilton thrown the full weight of her expertise behind a summer-long campaign to warn of the dangers of the new wine-colored gasoline, she might have sparked a forceful debate early, before Ethyl gasoline could become entrenched.

But there was more to Alice Hamilton's life in these days than workplace toxins and scientific journals. On June 21, 1923, Hamilton got word that Jane Addams, her friend and benefactor of more than twenty-five years at Hull House, had been hurt in Beijing. According to the telegram Hamilton received in San Francisco, Addams was in a rickshaw accident on a busy street and hurt her arm and her right side. She cut her trip short in China and left for Japan, but after several days she was still feeling tired and sore. She cabled Hamilton about the incident. Hamilton cabled back at once:

CAN COME IF NEEDED

And then several days later,

SAILING TWELFTH VANCOUVER EMPRESS RUSSIA

Hamilton dropped all prior engagements and took the RMS *Empress of Russia* from British Columbia to Yokohama. Addams was later diagnosed with a malignant breast tumor. Facing the possibility of death, Addams's call for Hamilton was a measure of friendship and a reflection of their shared sense of loneliness in the world. Both were unmarried in an era that stigmatized single women. Both had risen to the tops of fields mostly dominated by men. And both had grown accustomed to going through life with a hardened layer of puffed-up confidence to mask their inner sensitivity and insecurity. Merely the fact that Addams alerted Hamilton to her injury was enough for Hamilton to understand it was serious.

During Hamilton's week on the steamship, her reputation preceded her. A senior official in the Japanese government heard from a woman traveling with Addams that the renowned doctor Alice Hamilton was coming to Japan. When she stepped off the boat in Tokyo on July 23, a man on the dock proposed a schedule for her. After several days visiting and raising the spirits of Addams, Hamilton left on a month-long tour of Japan's textile mills and factories. She met with workers and conducted surveys and medical evaluations. Her findings came together in a lengthy report to the Japanese government to make its workers safer. She suggested better ventilation in factories, more durable uniforms for workers, and regular medical screenings for anyone who felt sick. Within two years the Japanese government implemented all of Hamilton's suggestions throughout the country.

During Hamilton's tour of Japan, a remarkable thing happened. A Japanese surgeon discovered that Addams's tumor was benign, and after a double mastectomy and several more weeks of rest, Addams began feeling better.

As the summer neared its end, Hamilton needed to return to Boston. She and Addams sailed together on the SS *President Cleveland*, a troopship taking a pause between the world wars to ferry Americans along what it advertised as the "sunshine belt to the Orient." Hamilton arrived well rested back on the East Coast in mid-September to a deluge of news and a pile of letters from fellow doctors, almost all of them begging her to join their public battle against General Motors—and, if she could, to use her intellect, experience, and credibility to fight back.

Midgley and Kettering were aware of the objectors. Of the dozens of letters doctors wrote to the surgeon general, many were carbon copied to General Motors, where no one knew what to do but to give them to Midgley. Midgley responded to many of them, stating that no experimental data had yet shown that Ethyl gasoline was in any way dangerous. But neither did he claim the opposite: that no experimental data had yet proven it safe.

Several GM engineers concocted a half-hearted study in the summer of 1923 at the research labs in Dayton. They ran a four-cylinder engine nonstop for one week and diverted the exhaust through a tank of distilled water to capture the metal dust. At the end of the test, there was hardly any dust to see. A week later they began another test to see how much lead would be deposited as floor dust. About fifteen cars in the lab's garage were filled with Ethyl gasoline and were run for thirty working days. The engineers relied on selective ignorance—looking only for large lead pieces and ignoring smaller airborne fragments—and reported they found nothing concerning. "The amount of lead found in the sweepings collected at the end of the test was so small that a person would have had to absorb each day all the lead from about 4.5 grams of floor dirt in order to be in danger of developing lead poisoning," a company report concluded.

Midgley performed a different test on his own using his

fingers. One day in the lab, he stuck his fingers into a beaker filled with tetraethyl lead and held them there for several minutes. For the next few days he experienced insomnia and loss of appetite. When a worker in his lab replicated the experiment, he was struck with "wild hallucinations of persecution." Midgley reported these episodes to a meeting of the American Chemical Society in December 1923. This was undeniable proof of toxicity. But Midgley spun it another way: He claimed he was simply identifying the threshold of toxicity. He argued that there were millions of harmless compounds in the world—like water, for example—that would become toxic at high enough exposure. Midgley's self-delusion was convincing. The ACS members nodded along.

Privately, however, Midgley knew that no quantity of internal studies and anecdotal research would silence the dissenting doctors. And so finally, in October 1923, Midgley and Kettering relented to a formal study of Ethyl gasoline. For a few weeks Midgley entertained the idea that Harvard should perform it. Alice Hamilton and her colleagues had recently undertaken deep investigations into lead's effect on the body; if anyone could give Ethyl a convincing exoneration, it was Hamilton and other Harvard researchers.

Midgley wrote to David Edsall, the dean of Harvard Medical School, and suggested it. The study would be "for the purpose of increasing the total of human knowledge and with no ulterior motive in mind whatsoever," Midgley wrote. General Motors would even pay for the study without any strings attached.

But before Edsall could consider the idea and consult any Harvard colleagues, Kettering ordered Midgley to snatch the offer back.

Kettering had a better idea. Harvard's scientists had already shown they were biased against lead. They would likely embark on a fishing expedition to find problems with Ethyl.

Instead, Kettering thought that General Motors should propose a scientific inquiry conducted by *the government*, which would not only go easier on Ethyl but had already proven to be reluctant to crack down on American business. If the U.S. Bureau of Mines performed the study, it would also have the credibility of Uncle Sam himself. When Kettering pitched the idea to the chief of the bureau, a meek geologist named Harry Foster Bain, Bain went for it.

It helped that Kettering offered for the General Motors Research Corporation to pay for the work *and* make its materials available to investigators. Kettering's goal, in other words, was to show the public his company had nothing to hide.

On its face, the deal looked sound. But there were details to work out. After Bain and Surgeon General Cumming agreed to Kettering's proposal and appointed investigators, Kettering made several small addenda to the agreement. In one, he wanted the investigators to keep all their findings secret until their final report. Releasing periodic updates was common in government research at the time, but Kettering feared it would create "scare headlines and false impressions" before the bottom line was clear. In another change, Kettering pushed that all official materials should refer to the substance in question as "Ethyl gasoline," and not "tetraethyl leaded gasoline." "Lead" was a loaded word that he feared could prejudice the study in the eyes of the public.

These were strange requests. For an official chemical investigation, it was uncommon for the government to use the brand name of a substance instead of its chemical name. Nor were government investigators accustomed to guarding preliminary results.

But looked at another way, these were minor points that didn't seem to interfere with the design or results of the investigation. So Cumming and Bain agreed. And once Kettering won these concessions, he asked for more.

Since the General Motors Research Corporation was paying for the study, he argued, his company should at least be
allowed to contribute. He added to the contract that, before
any publication, all research materials should be submitted
to GM for "comment and criticism." By this point, the bureau
had already begun to prepare for the study, and, again, seeing no influence over the actual scientific work, the government agreed to this new term too.

With everything else in writing, Kettering thought it
harmless two months later to add one final word. Kettering
asked that the part about "comment and criticism" be length-
ened to include "comment, criticism, and *approval.*" One
mere word seemed insignificant in an agreement composed
of thousands, and so, not wanting to upset a contract that
took months to write and derail a study that likely wouldn't
come to fruition otherwise, the government consented to
this as well.

Brick by brick, Kettering built a wall around Ethyl gasoline. He ensured there would be a government report that
would say everything he and Midgley wanted—and nothing
they didn't.

The Bureau of Mines' willingness to acquiesce to all of Kettering's demands was not a sign of corruption. Nor was its
rollover a symptom of fraternizing between government officials and their corporate friends. It merely reflected a government ill-prepared to confront a behemoth company that
was popular with the American public. The agency had no
room in its paltry yearly budget of barely $1 million to under-
take a deep investigation, and even if it did, U.S. patent law
protected Ethyl gas as a proprietary formula, which made
it almost impossible for the agency to produce test samples
identical to the ones refined, shipped, and sold. In other

words, a government inquiry into a privately produced substance required cooperation from the private company that produced it, and that cooperation came at a price.

The government investigation of tetraethyl leaded gasoline began on December 6, 1923, at the Bureau of Mines' research station in the small village of Bruceton, just south of Pittsburgh. The experiment station was created in 1910 atop an old coal mine for exactly this sort of work: to study properties of fuels like coal, petroleum, and gasoline. Bruceton was so proud to have the station in its community that several hundred men in top hats came in 1910 to see it dedicated. By luck, Bruceton was only a few hours' drive from Dayton.

According to the bureau's plan, the experiments would be conducted by five government scientists, all led by Royd Sayers, the head surgeon of the U.S. Bureau of Mines, who had previously studied odious gases like carbon monoxide, sulfur, and helium. Sayers was a government scientist, a thin man with a receding hairline frequently seen in a black suit, who opted for the prestige of government work over a more lucrative appointment with a private company. He was a government investigator tasked with protecting public health, but one who lacked the bold demeanor needed to push back on a corporation that tried to limit the duration and scope of his work.

Sayers and his assistants built a 1,000-cubic-foot gas chamber that occupied almost an entire room. To simulate the effects of breathing car exhaust, the chamber would pump air from an engine powered by Ethyl gasoline into one side, and on the other a large fan would suck the air out. The air would constantly be in motion, and he calculated that the air would completely circulate through every two minutes. If the exhaust remained any longer, the test subjects would be poisoned by carbon monoxide before they experienced any effects of lead poisoning.

For the next few months, the scientists placed a series of animals in cages in the chamber. There were forty-six rabbits, thirty guinea pigs, eleven dogs, and eight pigeons. Half the animals spent three hours per day breathing the exhaust in the chamber, and the other half spent six hours. The first increment was meant to simulate the average time the average person might be exposed to exhaust on a city street, while the second was to see if the six-hour animals accumulated twice as much lead or had symptoms twice as bad, which might suggest a proportional relationship of lead buildup over time. A third and smaller group of animals was kept in cages but unexposed to leaded gasoline. They would be the control group.

The animals weathered the exhaust with no visible symptoms. Their body weights stayed stable and they ate and digested food normally. One of the dogs even had a litter of puppies while inside the chamber. (Without missing a day, the puppies—named the Ethyl Gas hounds—were put into the experiment and returned to the chamber with their mother the next day.) It wasn't until the end of the experiments, when the animals were killed, that researchers could analyze them more closely. And when they did, they found broad signs of biological breakdowns. In nearly all the animals they found blood clots, swollen kidneys, distension of the heart, and "degeneration of liver, spleen, kidney, and adrenals."

This didn't necessarily point to lead as the culprit. The fact that the animals' lungs appeared to be almost uniformly normal at the time of death suggested that the inhaled lead hadn't had a fatal impact. And, strangely, the control group of rabbits also had signs of kidney, spleen, and adrenal swelling. With such murky findings, the researchers speculated that the animals' breakdowns were instead the result of other factors, like infectious diseases of closely confined animals or carbon monoxide in the exhaust. One wondered whether it was simply the rise in temperatures of the summer months. Amid so many stressors, it was impossible for anyone to say whether lead was the worst.

Phase two of the investigation would probe whether inhaled lead stayed in the body or whether it was exhaled back out without harm. This phase called for human test subjects who could sit still for long periods while wearing respirator masks. The test subjects, all men, were drawn from a pool of volunteers who believed there was nothing harmful about Ethyl gas. Lax ethics like this, in which people unknowingly signed up for substantial risk, were common at the time. It wasn't until 1974 that Congress required internal review boards to assess human danger in scientific experiments—and in many cases to forbid it.

In phase two, the male subjects were asked to wear masks that had a double valve system. A compressor would feed exhaust air in one tube, and a separate tube would capture the exhaled air. Once tabulated, this experiment presented striking results. A man who breathed in 0.180 mg of lead only breathed out 0.131 mg, indicating that his body—and specifically his lungs—retained a quarter of all the lead that entered it. Over thousands of breaths and dozens of years, one could extrapolate that the average person's lungs would retain hundreds if not thousands of milligrams of lead, enough to cause chronic lead poisoning. But the experiment also ran into constraints. After a few months of periodic exposure, the men showed no symptoms. There was no lead line on their gums that would indicate lead poisoning. Without killing the men and dissecting their bodies, it was impossible to know the precise state of their organs.

This led to an unsatisfying conclusion. Without evidence showing a clear cause-and-effect relationship between lead exposure and lead poisoning, the researchers had no choice but to conclude that the danger of lead poisoning from Ethyl gasoline was minimal at most, and their months of calculations failed to definitively show that a danger existed at all.

In hindsight, the bureau's finding laid bare the shortcomings of measuring a long-term problem in the blink of a few months. Symptoms of low-dose exposure tend to start slowly and increase with time. A brief government study could never measure the impacts of leaded gasoline over time. That would take decades. But it was reasonable to make a basic assumption: Routinely breathing in lead in gas form would have some negative effect, and doing it for decades would make it worse.

As it happened, the contract Kettering had meticulously negotiated with the Bureau of Mines to be favorable to General Motors turned out to be moot: As soon as Kettering caught word that the bureau's findings would effectively exonerate Ethyl gasoline without him having to meddle with the results, Kettering directed a GM colleague to write to a friend at the American Medical Association that the results would show "that there is no danger of acquiring lead poisoning through even prolonged exposure to exhaust gasses of cars using Ethyl Gas." And once the AMA disseminated this news to its members, Kettering's next move was to leak the results to the *New York Times*.

The Bureau of Mines study was a coup for Kettering and Midgley. And when it was published several months later, it might have been the final word on the safety of Ethyl gasoline.

If there was not a major crisis still to come.

# 10

## Elizabeth, New Jersey — 1924

After lunchtime on Tuesday, October 21, 1924, William McSweeney started feeling queasy at work. He was new on the job, so he didn't want to ask to go home. But by 3:00 p.m., he told his boss he needed to leave early. Thirty minutes later he was struggling to walk the two miles home to his house on Fulton Street in Elizabeth, New Jersey.

McSweeney was an immigrant from Europe. He was twenty-three and eager to fight for Ireland's freedom from the British. But when an opportunity came to move to America in late 1923 and live with his brother, McSweeney took the first ship he could find to New York City.

Within a week of arriving at Ellis Island, McSweeney was offered a job at Standard Oil of New Jersey's Bayway Refinery across the river from Manhattan. He was lucky to have work, people told him. So he felt excited on his first day. And he was probably extra excited when he was sent to a small redbrick building that, starting on August 1, 1924, had begun conducting a groundbreaking manufacturing experiment. When the other men joked that he was entering the "looney gas building," he laughed along with them.

Standard Oil of New Jersey was indeed performing a test. For more than a year, its competitor, DuPont, had been making tetraethyl lead using ethyl bromide. It was working at full capacity, operating at almost wartime production levels at the request of company president Irénée du Pont himself. "Every day saved means one day advantage of possible competition," du Pont told a colleague.

The competition he feared was from Standard Oil of New Jersey, where chemists thought they could make tetraethyl lead faster using ethyl chloride. Ethyl chloride was a bigger molecule that would form stronger bonds with carbon and thus have a higher boiling point, and it would not break down during engine combustion.

General Motors didn't care about the competition. It needed both companies working at full capacity, if not more. Based on the demands of gasoline distributors around the country, GM was expecting a fiftyfold increase in Ethyl sales in the first half of 1925, and Standard Oil of New Jersey (SONJ), sometimes referred to as Sohn-Jay, had considerable capacity. SONJ was one of the biggest players in American fuel refining even after the original Standard Oil—the company founded by John D. Rockefeller in 1870—was declared an illegal monopoly by the Supreme Court in 1911 and broken up into thirty-four smaller companies. In 1923, the General Motors Chemical Company came to partner with SONJ in a deal that awarded each company one half of the stock of a new company whose sole purpose would be to make, mix, and distribute Ethyl into gasoline tanks across America. It would be called the Ethyl Gasoline Corporation, or Ethyl Corp.

McSweeney's job at the SONJ plant was to mix a series of chemicals in a large vat that looked like two cones joined at their large end. No one told him what kind of chemicals he was mixing, but they were sodium, ethyl chloride, and lead. After mixing them, he and another man would remove a clear substance off the top of the container. This was the valuable material. The rest was a gray sludge that they'd dump through a grated drain in the floor. When the drain backed up, they'd push the sludge through with any pokers available, and if none were, they'd stomp on it with their boots or push it through with their hands. The sludge didn't stain or leave a residue on the skin, so the men rarely used gloves. For this they were paid an enviable eighty-five cents per hour, twenty cents more than laborers in the SONJ facility next door.

The hallucinations were a joke at first. Men who spent too much time with the gray sludge started seeing butterflies or erupted into fits of rage. Usually such symptoms would subside after a few hours or days. This was what McSweeney thought might be happening to him.

But the next morning his hallucinations were worse. His sister-in-law called the police and said McSweeney was shouting and acting irrationally. When a policeman arrived, McSweeney lunged at the officer with a knife. The officer called down to three men on the street to help subdue McSweeney and carry him, one holding each limb, to a hospital four blocks away, where he was put in a straitjacket.

Then things started to get strange. Two days after McSweeney's breakdown, another man working on the conjoined cone container became "suddenly delirious" on the job. Moments later he started convulsing violently and shouting that he was under attack by "three coming at me at once." Within minutes, he, too, was put in a straitjacket and hauled away.

A day after that, another man went home sick and awoke in the middle of the night so disturbed that he jumped out of his second-floor window. No one saw him jump or heard his fall, but a passerby noticed him several hours later moaning in a puddle of blood. He was rushed to the hospital and died the next day in violent convulsions.

Within a week, all three men were dead, two more were in grave peril, and all forty-three of the remaining workers at the Bayway plant were anxious they could be next.

Nothing like this had ever happened before in an American factory. Factory workers were accustomed to getting sick from their jobs and even dying. But the hallucinations and sudden violence added a new dimension. The head of the Bayway plant invited a doctor to inspect the remaining workers. A handful were found to have symptoms of lead poisoning and were sent home; almost all would suffer from long-term brain damage. Around the same time, medical examiners in New Jersey and New York learned about the incidents and opened investigations.

When the investigators started digging, they realized that this wasn't the first crisis associated with tetraethyl lead. Six months earlier, two men had died working with the substance in Dayton, and fifty employees of General Motors had been under observation for lead poisoning—one of whom was the famous inventor of tetraethyl leaded gasoline himself, Thomas Midgley.

But there was something different about the episode at Bayway: It occurred less than a mile from New York City, which happened to be the country's dominant media market. The New York newspapers were not only the most widely circulated broadsheets in the country; they also had the best and most aggressive reporters who could sniff out a story or scandal from their large networks of tipsters.

Within two days, someone at one of the hospitals tipped off journalists for the *New York Times* and the *New York World*, along with the *Sun*, the *Daily News*, and the *Brooklyn Daily Eagle*.

Overnight, the story grew legs.

The news of five men going insane in a New Jersey refinery made the front page of almost every New York newspaper the week of Monday, October 27, 1924. The story was a perfect fit for the era's sensationalist journalism, which valued shock over nuance and often rushed too-crazy-to-be-true stories to the printer with few confirmed details. New York was building its reputation not just as the country's biggest city but also as a power center of the nation's biggest issues. Reporting in the New York papers was often so authoritative that local stories were sometimes syndicated in smaller papers around the country.

Here was a story that had it all. The Bayway disaster was local to New York, it involved an issue of national significance, it was both crazy *and* true, and at its center was an unsolved mystery. With all these factors combined, the headlines that followed were strong enough to grab readers by the lapels and hold them for days.

ODD GAS KILLS ONE, MAKES FOUR INSANE, declared the *New York Times* on October 27.

MYSTERY GAS CRAZES 12 IN LABORATORY, the *Herald Tribune* reported.

The *New York World* followed a day later: GAS MADNESS STALKS PLANT; 2 DIE, 3 CRAZED.

The reporting had the breathless pace of a developing story. It divulged lurid details about how the men died to encourage readers to check in daily for updates. Each installment brought more questions than answers. Why had the men gone insane? What dangerous materials were they working with? How many others were exposed? And, most urgently, was the threat neutralized, or might it spread to the public?

In follow-up stories, reporters dug deep for sources. They talked to an undertaker who described the black-and-blue bruises on one of the corpses. They consulted a doctor who relayed the violent language spewed by one man as he died. And they spoke to the father of one of the victims who vowed vengeance on Standard Oil for killing his son.

Standard Oil, meanwhile, refused to disclose any details that might explain the men's behavior or even what the harmful substance was. The vacuum of reliable information sparked a chain of rumor and speculation. Some of the papers called the gas "ethylene," others "ethylchlorid." Both were wrong. Reporters sent queries to Standard's legal department at 25 Broadway, but the office had gone silent. Standard executives figured that if they didn't comment, the story might slowly run out of oxygen. But their silence had the opposite effect. Day after day, headlines about a "mystery gas" kept the story alive, like a giant puzzle every reader wanted to solve.

For clues, the reporters turned to medical experts for details about *how* the men died. A doctor at the Alexian Brothers Hospital who admitted one of the men spoke to the press. "From my observation of the case . . . I believe that the breathing of gas had gone on for a period of days and weeks, during which the poison had been gradually accumulated in the system," he said. "There apparently were no early symptoms, but as soon as the man's blood became saturated with the poison, the violent attack took place . . . it attacked the brain and nerves."

The doctors treating the patients also had few details to work with. They expressed frustration at SONJ executives for providing no information that might inform their course of treatment, such as whether to cleanse the infected men's adrenals, lungs, or brain. At Reconstruction Hospital in Manhattan, where many of the remaining thirty-five sick men were under observation, doctors relied on the men's descriptions of the chemicals—a little gray and the texture of mud—to decipher what they might have been working with, and in what quantities. Many of the doctors and other medical officials were "picking through the maze of heretofore contradictory assertions," reported the *Herald Tribune* on October 28.

One reporter stumbled upon a clue. Someone tipped him off that earlier that year, Midgley gave a presentation to the American Chemical Society during which he disclosed that "the dipping of one's finger into . . . tetraethyl lead brought on insomnia and loss of appetite." Midgley was reported to have said, "Its further seeping into the body produced wild hallucinations of persecution, the nature of which never varied." Those symptoms were eerily similar to what the men experienced at Bayway, which made tetraethyl lead the leading contender for the type of poison that killed the men.

The first Standard official to finally speak was the chief chemist of the Bayway plant, a man named Matthew Mann. Having worked in the plant, Mann was exposed to the same noxious fumes as the insane men, and as a result, he was suffering from a brain fog that delayed his thinking. When reporters for the *Times* and the *World* intercepted Mann going into the plant and yelled questions at him, Mann said he would go inside and deliberate about a statement. He returned fifteen minutes later and recited one sentence from a piece of paper.

"These men probably went insane because they worked too hard."

"Do they wear gas masks?" one of the reporters asked.

"Yes," Mann said. "They wear gas masks, which it is impossible for the gas to penetrate."

Were they properly trained?

They were, he said. But he added that the men apparently disregarded their training and put their lives at risk. (The father of one of the men refuted Mann's assertions: Not only did they not wear masks because they were not given masks, he said, they had been assured by their foreman that there was nothing harmful in the plant. "Otherwise [my son] would have quit," the father said.)

And with that, Standard Oil of New Jersey had no further comment.

Behind the scenes, however, Standard officials were frantic. They were in a "funk" of shock, a Standard executive remembered years later. The entire chain of command, up to Standard's president, Walter Teagle, was horrified that something so heinous had happened in their plant, on their watch, and that in barely three days it had mushroomed into national news. The company's leadership hardly knew any details about the Bayway experiment. The work on tetraethyl lead was a tiny piece of Standard's much larger chemical production work, and it had come with the implicit assurance of both General Motors and Du Pont. Standard's bosses did not ask questions about tetraethyl lead or its health-related side effects, because they didn't feel they needed to. Now, in a crisis they didn't foresee, they feared being liable for something they didn't fully understand—or, worse, that they could face criminal charges for negligence. For the moment, they decided that the safest route was to say nothing.

There was no mystery behind what toxic material the men were working with—at least, not in the mind of Yandell Henderson, the Yale physiologist, who knew immediately that it was tetraethyl lead. And in the absence of any official comment from Standard Oil, General Motors, or Ethyl Corp., Henderson had no hesitation speaking to reporters about the fuel's extreme risk to public health. He never missed a chance to mention that he had been warning about it for almost two years.

It was obviously tetraethyl lead, Henderson told the *New*

*York Times* on October 27, 1924. Not only was Standard Oil making this toxic substance, he pointed out, but its partner companies were selling gasoline containing it in several states. He regarded it as one of the greatest menaces to "life, health, and reason" and charged that it was "one of the most dangerous things in the country today."

"It has been known for several years that the mixture of tetraethyl lead would produce this effect," Henderson said. "The General Motors Company two years ago asked me to make a report on the substance. I reported that it was exceedingly dangerous to life and health, and that it would be criminal to impose so great a hazard on the general public." (An anonymous source at Standard responded that Henderson's claim was "bunk." No official would go on the record to elaborate.)

Henderson's statements to the press were half true. General Motors *had* asked Henderson—who had previously studied tetraethyl lead as a toxin—to conduct a study in 1922 on the physiological effects of Ethyl gasoline. Henderson even entertained the idea and asked whether Alice Hamilton might join him. But when Henderson declared that he would only take on a corporate funder if he were guaranteed full autonomy over the experiments and results, Kettering saw Henderson as a loose cannon who could not be directed, let alone controlled, and the idea was dropped.

But Henderson also revealed a deep misunderstanding at the heart of the debate over tetraethyl leaded gasoline. His comments made it sound as if the average American driver would be exposed to the concentrated sludge that poisoned the Bayway men and drove them insane. This was wrong. Motorists would encounter an extremely diluted form of it, on the order of a tenth of one ounce of tetraethyl lead per gallon of gasoline. Such a distinction made it much less dangerous to the public. When this distinction was later brought to Henderson's attention, he clarified that, even in diluted form, a tiny amount of lead dust would quickly grow into a big problem.

"If an automobile using this gasoline should have engine trouble along Fifth Avenue and release a quantity of gas with the lead mixture, it would be likely to cause gas poisoning and mania to persons along the avenue," he said. And even worse, he declared, since the gas had minimal odor and the symptoms of lead poisoning didn't occur immediately, a person could be poisoned anywhere in America without even knowing it.

To prove this point, Henderson thought up a simple and vivid demonstration that would quickly prove the damage of exposure to Ethyl gasoline. He bought five goats from a farm near New Haven. He shaved the animals and painted their bellies with Ethyl gasoline—not pure tetraethyl lead, but diluted Ethyl gasoline. Within a few weeks, all five goats died. Henderson provided no evidence that the goats hadn't died from other causes or that they hadn't been killed secretly to make a point, but the New York papers didn't wait to find out before reporting the experiment.

These explosive claims had the effect Henderson wanted. Within two weeks of the Bayway workers going insane, the board of health of the state of New Jersey shut down the Bayway plant and halted all manufacturing of tetraethyl lead in the state. Days later, New York City and Philadelphia followed suit and instituted their own bans on both the manufacture of tetraethyl lead and any sale of Ethyl gasoline.

For a moment, Ethyl gas seemed to be unraveling. A well-publicized public health disaster was exactly what Hamilton believed would disrupt Ethyl gas. And here, pure tetraethyl lead had violently killed five men with three dozen more under medical observation. As a public relations crisis, it was hard to imagine anything worse.

Now that the Bayway facility and several other plants were down, reporters turned to Charles Norris, the medical examiner of New York City, whose investigations would reveal exactly what the toxic substance had done to the men and what risk it posed to the bodies of everyone else.

Norris was the best toxicologist in New York. He was so good at performing autopsies that the mayor felt he had little choice but to hire Norris even after it was revealed that he once performed illegal autopsies to learn how to dissect a body.

At first, Norris wondered if he had jurisdiction over the Bayway victims. The poisonings technically occurred across the river, in New Jersey. But the victims had been rushed to, and died at, Reconstruction Hospital in Manhattan.

To Norris, the victims appeared to suffer from acute lead poisoning consistent with extreme exposure to *some* form of lead. But to know more than that, Norris summoned his deputy, a junior toxicologist named Alexander Gettler. Norris's and Gettler's names were often invoked in the context of medical mysteries in need of solving. And here was a true medical mystery.

It was also a rare opportunity to investigate lead poisoning in people rather than animals, as decades of prior researchers had done. Those earlier studies offered a portrait of how lead affects organs in mammals, but to know its precise effects on humans required human cadavers, which Norris and Gettler now had.

A week after McSweeney and the others died, Norris and Gettler started their autopsies in a small lab at Bellevue Hospital on First Avenue and Twenty-Eighth Street. They were permitted four bodies (the family of the fifth objected to the procedure) and began to dismantle them organ by organ.

The skin of all four men had turned yellow. Inside, they found burst blood vessels and engorgement of the spleens and kidneys. The brains had hemorrhages, and so did the lungs.

After making visual observations, Norris and Gettler conducted a chemical analysis to learn how much lead was in each corpse when the men died. They held each organ over a pot of steam to soften them, then collected small dust particles from the organs, which they mixed with sulfuric acid, sodium bicarbonate, and oxalic acid, which would each reveal lead in different forms. After spinning the samples in a centrifuge, they placed them under a microscope. The first sample showed needlelike crystals of lead sulfate. The second showed lead carbonate, and the third revealed lead oxalate.

This was confirmation that the men indeed died from an extreme amount of lead. When Norris and Gettler tabulated the results, all four men had at least 100 milligrams of lead encrusting their bones, livers, kidneys, lungs, and brains—more than fifty times the reasonable daily exposure for a person in that era. One man had more than 150 milligrams. "There was an astonishingly large quantity of the metal in the brain and body of [one] man, proving the high absorptiveness of the gas," Gettler told reporters. A man who had investigated some of New York City's most gruesome crimes admitted he was stunned by this case.

But the bigger finding was not just that lead could kill a man, which every doctor already knew. It was the impact of a certain type of lead. Unlike normal lead poisoning, which was known to target a person's veins, bones, and liver, these men had the most damage in other organs, leading Norris and Gettler to a sobering conclusion.

Tetraethyl lead was most destructive when it was inhaled. And from the lungs, it traveled straight to the brain.

# 11

## New York, New York — 1924

Midgley raced from Dayton to New York. The media circus had not subsided since the Bayway men went insane. In fact, with each day it seemed to be growing. In the five days since the first reports, reporters had put together more of the puzzle, and the horrifying details were now appearing in small newspapers across America. Every negative mention was chipping away at the already fragile reputation of Ethyl gasoline and its parent company, Ethyl Corp.

Such a crisis also threatened the reputations of three of America's biggest brand names—General Motors, Standard Oil, and DuPont—that were collectively worth hundreds of millions of dollars. And perhaps most acute for Midgley, his own credibility was on the line, along with the credibility of the hydrometers, valve systems, and everything else he had ever invented.

For the first few days, Midgley watched the media storm play out from his home in Dayton, where he read concerning news reports and the occasional telegram. Early scrutiny seemed to focus on Standard Oil of New Jersey and a mystery gas it was producing, which did not indict, or even mention, General Motors or Ethyl gasoline. But once the press started to link Midgley to the Bayway men due to Midgley's old speech to the American Chemical Society, he started to grow concerned. It wasn't until Midgley received a telegram from Kettering, who was traveling in Paris, that Midgley began thinking about damage control.

The story spread quickly and evolved as it went. Not only had the papers labeled Ethyl "looney gas," but Midgley felt that the sensational stories were causing millions of people to confuse the highly concentrated tetraethyl lead *gas* that poisoned the Bayway men with Ethyl *gasoline*, making people fear that if a drop of Ethyl gasoline accidentally spilled on their hands, their blood pressure would drop, they'd go unconscious, and they might even die before they could wipe it off.

Kettering rushed back from Paris to Dayton to help manage the crisis. More than feeling sadness over the men who had died or regret for the incident, Kettering was furious that a small episode at a single industrial manufacturing plant, of which there were thousands in America, had become front-page news across the country (or, as Kettering's boss Sloan put it gingerly: "Something like five men were lost and we received a very great deal of unfortunate publicity"). Kettering blamed Standard Oil of New Jersey and its foolish executives who had run such a shoddy operation that got out of control. Men had died at the Dayton facility and dozens more had gotten sick after mishandling tetraethyl lead. But those incidents had been dealt with quietly, away from the press. Now those episodes were being dredged up and connected to Bayway as proof of a much larger social threat that, in Midgley's and Kettering's eyes, was completely overblown.

A reporter met Kettering on his first day back at the office and asked him for comment: What did Kettering think about Yandell Henderson's goat demonstration?

"Doesn't that prove that your gas is poisonous, Mr. Kettering?" the reporter asked.

Kettering snapped back. "It proves only one thing, somebody has five dead goats!"

Kettering explained that goats painted with regular gasoline, without tetraethyl lead, might die too. After all, fuel could be lethal in any large exposure.

Kettering quickly realized that arguing with reporters didn't convey confidence in his product. So, instead, Kettering told Midgley they needed to calm the nerves of the public.

In Kettering's view, the key was to explain that raw tetraethyl lead might be poisonous in large doses, but it was perfectly fine in small quantities. And that Ethyl *gasoline*, which had been sold in thousands of gas stations for more than a year without any adverse report, had a minuscule, almost microscopic amount of tetraethyl lead, which made it completely safe. Besides, he thought, the world was full of substances that were known to be questionably or even obviously dangerous. There were the munitions and explosives made to end men's lives. There was mercury in thermometers, formaldehyde in textiles and furniture, and radium in cosmetics. The world hadn't gone hysterical about them. Their risks had been managed, and, for a time at least, they faded into the background of daily life.

This became the company's official strategy on October 30, 1924. After a week of silence, Standard Oil of New Jersey invited reporters to a press conference at its offices in New York.

When the reporters arrived, they saw Midgley standing on a stage behind a podium.

"Under proper safeguards," Midgley said, "the lead fumes from Ethyl gasoline are not deadly."

What about the main ingredient? one reporter asked. Was it dangerous to spill tetraethyl lead on one's hands?

Midgley asked an assistant to bring him a bottle of pure tetraethyl lead.

When he had it in his hands, he tipped the bottle and poured a thick amount on his palms and washed his hands with it. Midgley wore no gloves, and when finished, he rubbed the excess liquid off with a handkerchief.

"I'm not taking any chance," he said, "nor would I take any chance doing that every day."

To underscore his point, Midgley then held the bottle up to his nose and inhaled. He kept it there for thirty seconds.

"The fumes could have no such effect as was observed in the victims if inhaled for only a short time," Midgley told the room.

This was a shocking scene, considering the fresh memory of the men killed by the same substance, and would have been even more incredible if the reporters knew about Midgley's own lead poisoning one year earlier.

But, a reporter asked, wasn't it true that two other men died a year prior at another plant, in Dayton, that was making tetraethyl lead?

It was, Midgley said.

And weren't more than a dozen men hospitalized for lead poisoning?

They were, he said, declining to mention that he was one of the poisoned.

Then how can you say the substance is completely safe? someone asked.

On this point, Midgley didn't hesitate:

"This extremely dilute product has been for more than a year in public use in over 10,000 filling stations and garages and no ill effects thus far have been reported." He went on: "What injuries resulted were caused by heedlessness of the workers in failing to follow instructions."

When another reporter followed up wondering if Midgley was indeed blaming the workers, Midgley clarified that it was neither the substance nor the company behind it but the workmen, who should have known better about coming into contact with large quantities of it.

"Without desiring to attach any blame to the employees," he said while preparing to attach blame to the employees, "it has been found . . . that the men, regardless of warnings and provision for their protection, had failed to appreciate the dangers . . . of the fluid on their hands and arms." Whether he truly believed this is unclear, but he was committed to blaming anyone or anything to protect Ethyl.

A representative for Standard spoke after Midgley and underscored this point, crafted conveniently to absolve the company of any wrongdoing and instead lay blame on a few bad apples.

The men were not working in a post office or a flower store, he pointed out. They should have known when walking through the door of a chemical manufacturing plant that they were engaged in "a man's undertaking." The message seemed to be clear that if these men had been a little smarter, more educated, and had spoken better English, the incident would not have happened.

This was a compelling argument in a time when workers' lives were already widely devalued. And Midgley's demonstrations with his hands seemed to be convincing proof of his claim that the harm of small quantities was overblown.

Scientists and historians have scrutinized this episode and wondered whether Midgley truly poured liquid tetraethyl lead on his hands at the demonstration or whether it was something tamer with the same appearance and consistency, such as glycerin. But no one asked and it was never proven.

After an hour, the questions petered out. If Midgley had pulled a ruse, it appeared to have worked.

The press conference had its intended effect. The journalists returned to their bureaus and published stories the next day recounting Midgley's demonstration and his responses to their questions.

But it would take more for Midgley to stop the downward spiral of Ethyl and all the companies now invested in its success. The firestorm was still rippling across the country. The New Jersey Board of Health had closed the Bayway plant, and Philadelphia and New York had halted the distribution and sale of Ethyl gasoline, but even more serious implications were underway. Labor unions in the Northeast were proposing a national ban on Ethyl gas. In New Jersey, the attorney general assembled a grand jury to weigh whether to bring criminal charges against Standard officials, from those at the top of the company all the way to the Bayway plant foreman, and if the jurors recommended indictments, then criminal trials would keep the story alive for months, maybe years. Across America, the controversy was causing distributors to jump ship and disavow any relationship with Ethyl gasoline. The Gulf Refining Company, one of the largest gasoline refiners that owned more than 1,000 filling stations across the South, ran a large newspaper advertisement announcing it would immediately stop distributing Ethyl gas "until it can be demonstrated beyond question that it is not detrimental to the public health."

This was a crisis, and fighting back called for a new way of communicating with the public. The practice of public relations was barely a decade old, and there had never been a corporate snafu this big, this far-reaching, and with so much money at stake. The controversy called for the best public relations experts in the country, if not *the* best one, who happened to be a forty-seven-year-old former journalist from Georgia named Ivy Lee.

As a young man, Lee was working as a reporter when he noticed that he could make more money helping protect big companies from harsh reporting. His first big chance came in 1906 when he helped the Pennsylvania Railroad Company recover its reputation after a train fell off a drawbridge in Atlantic City and killed fifty people. Seeing how the story and the assigning of blame could get whipped up in the media, Lee pioneered a way to steer the narrative by communicating in a way that seemed transparent and honest but also withheld unflattering information. He went to the scene of the accident and drafted a favorable news report with detailed information and quotes from law enforcement. Then he delivered the document he labeled a "press release" to the offices of every New York paper. The next day, the *New York Times* published Lee's dispatch word for word.

By luck, Lee was now on the payroll of Standard Oil and its dozens of sibling companies. Standard's founder, John D. Rockefeller, hired Lee personally in 1914 to "burnish the family image" of the Rockefellers after the company was linked to a deadly coal mine strike in Colorado. The incident risked embarrassment for the world's richest man—and could possibly cost him money too. Lee went to work immediately, not just by distancing the Rockefellers from the Colorado story but also by humanizing his boss as a man who cared deeply about the common worker, emphasizing Rockefeller's generous donations to fund civic projects like hospitals and parks. Lee invited reporters into Rockefeller's New York and Florida estates to see his full gregarious generosity on display while he hugged his grandchildren and gave dimes to poor children. It was the earliest form of corporate whitewashing—or, as it would become known, "crisis PR." Put another way, it was the first demonstration of a form of reputational laundering available to those wealthy enough to afford it. Most valuable of all for Lee, however, was that his success earned him the trust of all Standard Oil companies and rendered him a bit like a corporate ambulance, on call in an emergency.

Lee met with Midgley after the press conference at Standard headquarters. To fight back against the boycott ad from the Gulf Refining Company, Lee advised Midgley that Ethyl should create an advertisement of his own. In it, he had to underscore the difference between tetraethyl lead and Ethyl gasoline. To make the distinction clear, Lee suggested that Midgley should admit that tetraethyl lead was indeed "a poison" and had been responsible for the men's deaths in New Jersey. Putting this blame on an industrial ingredient most Americans did not understand and would not encounter in large quantities would convey truthfulness, which would build credibility when claiming that Ethyl *gasoline*, by comparison, was not only safe but was the only way to conserve the world's gasoline supply.

Midgley and Lee wrote the ad, opting not for pithy slogans but beefy paragraphs that they thought would convey that they were being forthright and comprehensive. They claimed that 20,000 filling stations were distributing Ethyl gasoline and that, since 1922, motorists had bought more than 200 million gallons of the fuel. The numbers were generously rounded up.

Lee placed the ad in all the New York papers. But, seeking a wider reach, he suggested that the advertisement be mailed to the hundreds of regional distributors of Ethyl gasoline with the request that they publish it in a paper in their territory at Ethyl Corp.'s expense. Just send us the bill, he told them.

Every recipient of Midgley and Lee's ad, from the major companies that ran filling stations to the mom-and-pop pump owners, had little reason to balk at the messaging. Ethyl's corporate crisis was becoming a local crisis, too, as customers began to pepper pump owners with concerned questions about the fuel—or, worse, they'd fill their cars at competing filling stations that didn't carry Ethyl. Anything that would help calm their customers, particularly the ones paying extra for premium gasoline, was welcome.

By and large, the effort worked. Starting on November 7, 1924, barely two weeks after the first man went insane in New Jersey, Americans awoke to gentle assurances in their local circulars.

# Ethylized Gasoline

For one year and nine months ethylized gasoline has been on sale. It is now being distributed through about 20,000 filling stations covering one-third of the territory of the United States.

About 200,000,000 gallons have been used by more than 1,000,000 motorists with complete safety and satisfaction.

Recently a distressing accident occurred at an experimental plant, where a new process for the manufacture of tetraethyl lead, one of the constituents used in ethylizing gasoline, was under development.

Tetraethyl lead is a poison, as are many raw materials which enter into the manufacture of harmless compounds. Ethylized gasoline consists of 1,300 parts of ordinary gasoline containing less than one part of tetraethyl lead.

This statement is issued to make plain the all-important difference between tetraethyl lead, the raw material, and ethylized gasoline, the commercial product.

Ethylized gasoline is more than an improved fuel, giving smoothness to the motor and eliminating knocks; it is a scientific discovery which, in its ultimate development, will contribute largely to the conservation of the world's supply of gasoline.

Exhaustive tests have been conducted which have established the safety of ethylized gasoline when used as a motor fuel. These tests have been confirmed by the United States Bureau of Mines, which is making additional studies to determine whether any possible injury can result from continued contact when used for other than motor purposes. Scientific data based on these studies will be submitted to any health commissioner or other public health official upon request.

## Ethyl Gasoline Corporation
### 25 Broadway—New York City

The message was clear: The controversy in New Jersey they read about was the result of a handful of poorly trained workers and was overblown by New York media muckrakers. Everyone could relax, there was no threat to the public, and the U.S. Bureau of Mines would issue a full report in a few weeks putting to rest any additional concerns about Ethyl gasoline.

One imagines the confusion of the average farmer, shopkeeper, or junior high school teacher who woke up and found Midgley and Lee's ad in their copy of the *Chattanooga News*, the *Baltimore Sun*, the *Cincinnati Enquirer*, or more than 150 other newspapers. In the prior weeks, the public had read about a horrifying disaster linked to an obviously poisonous substance. And now here, in front of their eyes, was the assurance of a respectable American company headquartered on one of America's most prestigious streets explaining *in detail* how the events were just a small misunderstanding.

A discerning reader might have been skeptical. The 1920s had been full of stories about inhumane working conditions, bloated corporations, and the chummy relationship between companies and the government agencies that regulated them. The Ethyl story embodied all these themes.

And yet the advertisement had the mollifying effect that Lee thought it would. Most Americans were not discerning readers—at least, not about an event that had not occurred in their community. They had soil to till, stores to manage, and seventh-grade biology classes to teach. Most people who could muster concern about the direction of the country were focused on bigger issues, like the recent Great War that had killed more than 100,000 American soldiers, or the cultural rebellions in music, dance, and fashion that were threatening traditional values across America. Compared to that, gasoline was boring.

This was terrifically helpful to Midgley. And in the weeks following Bayway, he made his way to the health boards of

Philadelphia, New York, and several other cities that had banned or paused the use of Ethyl gas, often with the PR chief Lee sitting at his side. Midgley didn't need to convince health officials not to worry about health matters. All he had to do was persuade them to wait for the Bureau of Mines report on Ethyl gas, which would be published any day. With the imprimatur of the government, he hoped the study would convince everyone involved in the making, shipping, and consumption of Ethyl gas that the worst had passed and there was nothing more to be worried about.

On December 1, 1924, the Bureau of Mines sent the media its Ethyl-approved report about the experiments with animals that it had conducted earlier that year. The document was brief, just twenty-three pages, and it detailed the way the animals at the Bruceton research center had been locked in cages and forced to breathe Ethyl exhaust. There were tables and graphs showing that there *was* danger to the animals and that lead *had* accumulated in their tissue after just a few days. This suggested danger for garage workers and factory men handling the substance, at least until better safety measures could be found to make handling concentrated quantities of tetraethyl lead safer.

But the bigger question, the one most readers of daily newspapers wanted answered, was whether there was harm to the general public. Was there evidence that breathing the exhaust of a car running on Ethyl gas could bring the same effects the Bayway men had suffered onto the average person walking down West Main Street in Louisville or Auburn Avenue in Atlanta?

Journalists speed-read the report and leafed through the bureau's findings looking for a conclusion. Until finally their eyes were drawn to one word on the final page. The danger to a person on a sidewalk getting lead poisoning from breathing the discharge of automobiles running on Ethyl gasoline was, in a word, "remote."

"Remote" was a curiously vague word in such a context, the kind that appeared to carry meaning but could be blown over with a single breath. "Remote" lacked any sense of scale. The north pole was remote, but so was Jupiter. Did a remote risk to the public mean a slight risk? Or no risk? As a conclusion, "remote" was squishy. The Bureau of Mines hadn't measured the danger of Ethyl gas to the average American. For all the bureau's scientists knew, the risk might be significant, or it might start slow and build over time. It might even cause people to die. But the data on a few dozen animals exposed to exhaust for a few months seemed to suggest that it was not unacceptably dangerous, which implied to the casual observer that it was, in effect, safe. With such a hazy finding, the bureau was effectively concluding that its conclusion was, in fact, inconclusive.

Journalists had little time for hedgy language. They wanted firm statements that could become firm headlines. If they weren't given a conclusion, their editors would make one themselves.

The next day, the *New York Times* cut to the chase and reported something the bureau's scientists had never said, never found, and likely didn't mean. But it was the best headline that Kettering, Midgley, and Ethyl Corporation could have hoped for.

## NO PERIL TO PUBLIC SEEN IN ETHYL GAS

*Bureau of Mines Reports After Long Experiments
with Motor Exhausts*

*More Deaths Are Unlikely*

It was the vindication needed to slow the bleeding. Standard's facilities were still shut down. New York and Philadelphia still kept their ban on Ethyl gasoline. And New York papers were still running follow-up stories about the Bayway men and their families. But there was a path back for Ethyl. So long as no one else uncovered any more undesirable evidence, it appeared for the moment that the worst had passed.

# 12

## Cambridge, Massachusetts — 1924

Before the Bayway incident, Hamilton had misgivings about fighting publicly against Ethyl. Tetraethyl lead was clearly a menace, yet so were dozens of other industrial chemicals that she had studied in her overstuffed files. And besides, was she really willing to risk her hard-won credibility on a substance that was so popular with the American consumer? But her opinion changed when she learned what happened to the men at Bayway and, even more horrifying, how the corporate bosses responded.

When the Bayway deaths occurred in the fall of 1924, Hamilton was in Moscow at the invitation of the Soviet Union's Department of Health. Moscow was dreary and cold that October, a metaphor for Russia broadly. The country was trying to move on from years of civil war and famine and was struggling to get people housed and fed. The Kremlin invited government officials and business experts from Europe and the United States to visit and advise on how to build "responsible industries," they told her. With the invitation to Hamilton, the question was how to do it safely.

Russia's eagerness to advance had a side effect that Hamilton noticed instantly. The country seemed to need every pair of hands to move itself forward. "The women do not waste time over housework and children," Hamilton observed in a letter to her family. "[The government's] ideal is to free the woman as much as the man." Starting in 1917, the Soviet Union empowered women to vote, legalized abortion, and pushed them to join the workforce, regardless of the job or industry. On almost all matters of gender progressiveness, Russia outpaced the United States. Despite the dank guesthouses she stayed in, the stale breakfast bread she ate each morning, and the bland coffee thrust at her throughout the day, Hamilton found this part of Russia fresh and exciting.

While in Moscow, a person in her traveling party received a letter from home that included the news from Bayway. It had become a "media storm," the correspondent wrote, that had overtaken the *New York World* and the *New York Times.*

This news likely filled Hamilton with both disgust and delight. On the one hand, here was sad news of another preventable industrial accident. But on the other, it seemed, remarkably, to finally have gotten the attention of the public in the country's biggest newspapers.

Hamilton knew enough about tetraethyl lead to understand what had happened. She scribbled the symptoms of the Bayway victims in one of her pocket notebooks as though collecting data on yet another poisonous substance. "This is a very dangerous form of lead," she later wrote in a more polished paper on the substance. "It is more quickly absorbed than any of those ordinarily used in industry and concentrates in the central nervous system, causing insomnia, excitement, twitching muscles, hallucinations like those of delirium tremens, even maniacal attacks and convulsions, and death."

She arrived home from Russia two days after Thanksgiving to a pile of papers. There were letters needing responses, reports needing review, and calls that had come in weeks prior. Buried in the pile were newspaper clippings from Henderson at Yale and Edsall at Harvard.

It was a lot for Hamilton to digest all at once. Five men had died from tetraethyl lead exposure. But Hamilton also read about other incidents from the last two years that were just now coming to light. In 1923, a worker at a DuPont facility in Carneys Point, New Jersey, died after handling the substance, and then ten months later two men died in Dayton at the General Motors Research Corporation labs. Two more men died at DuPont plants in 1924, one in July and one in October, a week before the Bayway episode. What was more, there seemed to be clear evidence that dozens of other men who came in contact with tetraethyl lead experienced some health impact—from high blood pressure to intense abdominal pain—on their way toward more extreme symptoms.

Looked at one way, these sorts of details had become run-of-the-mill in Hamilton's view. She was now fifty-five. It had been twenty-seven years, half her life, since she moved into Hull House as a young doctor and first observed the misery inflicted on the poorest and most marginalized people in society. In 1924, news that several dozen men had gotten sick or died, even violently, differed little from what she had seen in 1911 when she observed lead workers in Chicago.

But in another sense, Hamilton was freshly horrified by what had happened at Bayway—not necessarily about the deaths themselves but what followed. For one thing, the explosion of news coverage startled her. Never before had she seen such dogged and relentless reporting on an industrial accident. Were New York reporters aware that factory mishaps happened almost every day in America? Or was it the gripping details of the incident—the men in straitjackets, the man who jumped off the balcony, the days-long mystery of identifying the toxic substance—that made this story so sensational?

And there was something else. The way Ethyl Corporation and its partner companies were fighting back. Between the New York press conference and the newspaper ads, the companies seemed to be engaged not in gestures of humility and regret, however shallow, but in a brazen reputational offensive. In Hamilton's experience, after she pointed out a clear shortcoming, such as men needing proper gloves or masks, the factory owner would usually humbly agree to purchase the materials and send her on her way. Blaming lowly workers for plant accidents was not new, but waving away any sense of responsibility or remorse surprised her, especially when she read about Midgley's press conference.

She was also surprised to see the Bureau of Mines study that so completely exonerated Ethyl gas. Why hadn't the researchers done a more comprehensive study or drawn a sharper conclusion? It seemed obvious to her that Ethyl had not only funded the study but helped design it, which made it both shocking and yet entirely unsurprising that the bottom line was flimsy enough for Ethyl to exploit in its public relations. Ethyl's funding of the study "cast[s] doubt on negative results obtained by the investigators," she later wrote in a letter to Surgeon General Cumming.

And that was the last thing. The expensive campaign to calm the public was far more extensive than any corporate response she had ever seen. A coordinated ad blitz in more than a hundred different newspapers cost thousands of dollars, which suggested Ethyl was willing to put up whatever it cost to protect its product. Whether or not Ethyl Corp. believed Ethyl gasoline was as safe as it was claiming in public, it clearly believed it was lucrative. That alone seemed to explain the scientific gobbledygook spouted by the company.

Among Hamilton's Harvard colleagues, the most outraged

was a young researcher named Paul Reznikoff. Reznikoff, just twenty-eight, was a doctoral fellow who had gotten his MD three years prior from Cornell University in the science of blood, although his real interest was blood poisoning. When Reznikoff read the news reports about the Bureau of Mines study with all the animals, he was at first confused. Then he grew skeptical. Something seemed off with the data and the research, which seemed to wildly skew the conclusion. Eventually he grew steaming mad, and the best person a young man deeply passionate about blood could think of writing to was someone deeply knowledgeable about lead.

Reznikoff wasn't interested in challenging Ethyl Corp. He was a bookish blood scientist wound up by experimental data. Wasn't it weird, he asked Hamilton, how few animals went through the tests? Just over fifty. And the mortality rates were extremely high: 50 percent of the rabbits and 25 percent of the guinea pigs died during the tests. The researchers blamed the deaths on infections and accidents that affected the control group, too, which suggested the deaths were unrelated to lead. But Reznikoff thought this was a copout. Why had the bureau started with such a sickly and small population of animals for a question as important as this?

Hamilton was intrigued by these questions. Everything Reznikoff had brought up was a valid critique of how the experiments were conducted. Was this a case of laziness by government scientists, or was it corporate misconduct trying to influence an independent study? Hamilton wrote back to Reznikoff to meet her at Harvard the following Monday and they could craft a plan to fight back.

It was raining on December 8, 1924, when Hamilton settled into her Harvard office to meet Reznikoff. A storm had been inching up the East Coast and would soon drop snow for three days in Cambridge and bring the medical school to a crawl. Hamilton's thirty-minute walk from her brownstone in Back Bay to the medical campus at Longwood could be doubled, even tripled, if the sidewalks weren't shoveled, and they rarely were.

Tetraethyl lead seemed to be all anyone was talking about at the medical school, especially among her close colleagues. She overheard a conversation between David Edsall, the dean, and her physiology colleague Cecil Drinker, the future dean who would go on to establish Harvard's school of public health. After reading the Bureau of Mines study, the two men had come to the same conclusion as Reznikoff: The study was weak in design and execution, and it collapsed under basic scrutiny. The two had wound each other up over the high death rates of the animals being waved off as "accidents." They were also perplexed about the air recycling every two minutes. Nowhere on earth did anyone breathe recycled air every two minutes, especially not in a city where automobile fumes lingered in the streets for days.

She told the men that Paul Reznikoff had seen similar shortcomings and was on his way to her office to concoct a strategy to criticize the study and, with luck, force a new one.

Edsall enjoyed that his friends seemed united on this. He proposed a friendly competition to see who could write a critique of the study and get it published first. Hamilton knew better than to take any bets, particularly with Edsall, who was good-natured but competitive and also, incidentally, the editor of the *Journal of Industrial Hygiene*. Sure enough, two months later, Edsall would rush a brief critique into the February 1925 edition of the journal. "As an investigation of an important problem in public health in which a great many inexact data have already appeared," Edsall wrote, "[the Bureau of Mines study] is inadequate."

Reznikoff walked into Hamilton's office that day with the

energy of a young doctor eager to get to work. At twenty-eight, Reznikoff was the same age that Hamilton had been when she moved into Hull House as an optimistic and enthusiastic new doctor. Hamilton was "pleased and collegial," in her words, that Reznikoff had come to her for advice and help.

Reznikoff had a stack of papers and showed Hamilton his calculations about how much faster carbon monoxide poisoning happens than lead poisoning and how the measurements of lead dust undercounted the full lead exposure.

"The amount of lead was kept extraordinarily low in the air of the test chamber," Reznikoff told her.

Hamilton looked at his numbers. He was right. If the level of lead in the tests was anywhere close to what a normal person could expect to inhale in a garage or beside a busy street, the damage would be much worse. And it was easy to extrapolate how, over time and over millions of gallons of fuel burned, tetraethyl lead would become a poisonous plague to every living and breathing person in America, if not the world. There was no outrunning it.

Hamilton told Reznikoff that they needed to get to work. Their first task was to write a letter to the nation's top health official, Surgeon General Cumming. And then, she said, they would have to marshal the force of reasonable doctors everywhere to make the factual argument that no amount of caution was sufficient when talking about an obvious poison.

Hamilton was not a natural fighter. She had gotten by for more than five decades with a demeanor of persistent civility. But Reznikoff had underscored an important point. And after he left her office, Hamilton experienced an emotion rare for her. She felt angry.

Hamilton was fed up. She was also tired. Not only was she still recuperating from her monthlong travels in Russia. She also felt pulled between other industrial poisons that seemed to require her every waking moment.

The 1920s were a busy time for corporations trying new things with new materials. Almost every day there was news of a new chemical material that was too exciting to fully investigate before it was released to the public. A material known as Bakelite was yielding new advances in the field of polymer chemistry. (It would soon help make the first plastics.) A synthetic fiber known as rayon was gaining popularity in textiles due to its affordability and versatility. And adding chromium to steel made a new metal that was rustless—or, as it was later called, "stainless"—which grew into a booming industry. Advances like these put pressure on Hamilton, Edsall, and Drinker, who were tasked, formally or informally, with researching the health impacts of what felt like every new material.

In the winter of 1924, Hamilton was pulled into a new episode of corporate negligence resulting from *another* industrial poison. The call for her urgent attention signaled that her field had expanded quickly and that she was still the head of it.

The U.S. Radium Corporation, headquartered in New York, had grown in popularity since it debuted glow-in-the-dark paints in 1917. The glow was caused by radium, an element poorly understood but quite beautiful on a watch dial. In its main plant in Orange, New Jersey, the company employed several hundred women to paint luminous watch dials by hand. Like thousands of workers Hamilton had encountered before, the women were given no warning of the danger of their materials, and, as a result, they did not question their bosses when instructed to dip their paint brushes between their lips to sharpen the points.

Radium as a substance is profoundly damaging to the human body, a fact the Radium Corporation bosses selectively misunderstood at first and then deliberately ignored. Like lead, it exists in the earth's crust and can be naturally encountered in small quantities. But unlike lead, radium is radioactive and emits alpha particles that attack cells and DNA.

Sure enough, the dial-painting women started to show the predictable effects of radium poisoning. Their teeth and hair fell out, their joints swelled, and they developed lethal anemia. By 1925, after at least three women had died and hundreds more were near death or in debilitating pain, a large group of workers and former workers sued the Radium Corporation on the grounds that the company knowingly exposed its employees to a terrible poison and concealed its dangers for years.

This was the moment when the New Jersey Consumers League asked Hamilton to look into the case. She spent several weeks reviewing every aspect of a prior medical investigation of the affected workers. Her research left little doubt that the women's illnesses were the result of radium poisoning, that the factory should be shut down, and that the workers ought to be compensated for their injuries.

Even stranger, though, was how the radium episode seemed to mirror what was happening with Ethyl gas around the same time. A corporation had discovered an innovative product. It marketed it successfully to an enthusiastic public. And when victims started to surface with obvious health problems, the corporations didn't change course. They doubled down and asserted that their methods were sound and their detractors were somehow mistaken.

In addition to Hamilton's interviews with the women, Drinker at Harvard conducted a more granular analysis of the effects of radium poisoning on the women's bodies. Little about the investigations came as a shock to the two of them. But it did surprise Hamilton when Arthur Roeder, the president of the U.S. Radium Corporation, went so far as to falsify Drinker's final report about the radium girls, attributing to Drinker an incorrect finding that exonerated his company and pronounced that "every girl [was] in perfect condition."

"Do you suppose Roeder could do such a thing as to issue a forged report in your name?" she wrote to Drinker.

But she already knew the answer. She was wise to this new trend in American business, the way corporations would go to great lengths to protect themselves by introducing doubt and outright lies, even when the evidence was piling up against them. Detractors who wanted to litigate boring matters like details and ethics risked getting drowned out by the excitement of innovation.

All around her there seemed to be evidence of companies not operating by facts and reason but playing solely to the court of public opinion, where lavish marketing budgets could spin a favorable narrative. And if that didn't work, the organizations would simply resort to lawsuits, smears, or false accusations to silence their critics. It was an open secret that the Ford Motor Company used strong-arm tactics to crush labor unions and that the American Tobacco Company used dark campaigns to discredit doctors concerned about the health risks of tobacco. The same would soon happen with the Johns Manville company, which ferociously defended its proprietary insulation, known as asbestos, even after workers started to die from lung diseases.

But there was something about Ethyl Corporation that bothered Hamilton even more. Its product was not novel or experimental. It was lead, plain and simple, the same metal considered poisonous for thousands of years. Except Ethyl transformed lead into an even more dangerous compound that was easily manipulated, diluted, inhaled, and absorbed. Five years had passed since Midgley first demonstrated tetraethyl lead's anti-knock power in gasoline, and since then

Hamilton had not seen a shred of evidence suggesting it was harmless to humans.

Writing to Cumming was not enough. She had to wage a bigger battle against the men of Ethyl Corporation, who she knew had big egos and would dismiss her opinions. And she learned from Henderson that no amount of criticism could silence the deep-pocketed PR machines of General Motors, DuPont, and Standard Oil.

But she did have someone in mind who could help. On December 16, 1924, Hamilton sat down to write to Walter Lippmann, her old friend from the House of Truth in Washington. Since those days, Lippmann had risen to editorial page editor of the powerful *New York World*. If Ethyl couldn't be shut down by a shallow government study or critiques by scientists in journals, then perhaps Lippmann, a master of marshaling intellectual thought into public outrage, might have other ideas.

Hugh Cumming was in an odd spot for the quiet job of surgeon general. He was at the center of what was becoming a national melodrama. The Ethyl controversy had all the makings of a gripping trial: characters, tension, high stakes, and surprising twists. And a week before Christmas 1924, Cumming found himself as judge, juror, and potential executioner.

Cumming was unaccustomed to such controversy and did not enjoy being the center of attention. The office of surgeon general was not merely ceremonial, but it was also not a traditional hotbed of power. The role was an awkward holdover from colonial days when the only public health the government could reasonably influence was military health, which required that the country's top health official be a military officer. In the century and a half since the American Revolution, the country's population had ballooned by more than

100 million, its border had reached another ocean, and it had added thirty-five more states. But the office of surgeon general stayed largely the same, along with its odd military requirement. Cumming's experience in the Navy qualified him for the post, to which President Woodrow Wilson appointed him in 1920.

Conflict was also not Cumming's persona. He was tall and thin, like the captain of a rowing team, and wore round spectacles that framed his wavy white hair parted down the middle. He was a Washington power broker whom *Time* magazine described as charming and soft-spoken, and—even rarer in Washington—he seemed to have no enemies. "Few men are better known in Washington," the magazine reported. "When a telephone rings and his soft voice asks something for his Public Health Service, he gets that something very quickly."

The job of the Public Health Service, and Cumming's duty as its leader, was to oversee a network of aging military hospitals, along with America's response to contagious diseases like cholera, polio, and influenza. In 1912, the service expanded its scope to also cover noncommunicable diseases, which eventually put Cumming in charge of a long list of public maladies, including lung diseases, cancer, and diabetes. This enlarged his domain from simple disease control to regulating the broad effects of industrialization, including pollution, poor working conditions, and unhealthy living environments.

Despite his broad purview, Cumming had not been the one to initiate the meeting now happening in his office. Almost two months had passed since the Bayway meltdown, and Ethyl remained on shaky ground. Midgley stunted the hysteria with his and Ivy Lee's clever public relations. But despite the ongoing local bans, the worst damage seemed to be coming from within. After the breathless news coverage died down, the board of Ethyl Corporation was still furious at such an embarrassing public disaster. The incident had cost them $100,000 in settlement payments to the victims' families. And that was a small fraction compared to the bigger reputational cost. Bayway was a black eye for three major American companies that had significantly bigger interests than a small experimental brand of gasoline. If the other companies had to cut Ethyl loose to save themselves, they would. In the winter between 1924 and 1925, the Ethyl board debated whether the company should shut down Ethyl temporarily—or permanently.

Kettering lobbied to delay any decision about Ethyl's future. He told the board that he and Midgley needed time to rebuild Ethyl's reputation and trust. They would do this by finding a new way to manufacture tetraethyl lead that posed no risk of exposure to any worker. They would also secure greater confirmation than the Bureau of Mines study, which had been widely panned by Hamilton and others, that Ethyl gasoline posed no danger to the general public.

In the end, the board agreed and gave the men six months to build Ethyl back or kill it off.

That ultimatum was what brought the Ethyl men to Cumming's office. It was Christmas Eve 1924, a Wednesday, and most of the staff of the Public Health Service had left early for the holiday. Cumming sat behind his desk in his wood-paneled office on the fourth floor of the agency's headquarters a block east of the U.S. Capitol.

On the other side of his desk sat Charles Kettering, Frank Howard of Standard Oil, and an engineer for DuPont named Willis Harrington. The three men had come to ask Cumming for a favor.

It was Kettering's idea to see Cumming. He was on a tour of Washington to enlist public officials to sign off on Ethyl gasoline in order to put the matter finally to rest. He had already met with the secretary of the Commerce Department, a civil servant named Herbert Hoover, who had taken an interest in anti-knock fuels. The contents of Kettering and Hoover's conversation were never written down, but after their meeting, Hoover, a man clearly interested in rising to higher office, dropped any public mention of Ethyl that could be construed as criticism.

Kettering had a different request for Cumming. He wanted the Public Health Service to hold a public hearing about tetraethyl lead. Such a hearing would publicize the immense evidence Kettering believed he and Midgley had amassed about the fuel and its safety. A hearing would also put an end to the monthslong media controversy, offering a chance for all parties to get together and air their disagreements. Kettering believed that if he could flood the debate with excuses about past incidents, then the hearing's conclusion—that the worst of Ethyl gasoline had passed—would be a powerful endorsement for its future. And if he didn't think of one other factor, he soon would: A sign-off from the nation's top public health agency would help sell the fuel overseas.

Cumming considered the request. The four men discussed what a hearing would entail. Cumming asked for a list of experts who might be invited to testify. And then he said he would think about it.

Ultimately, it was an easy choice. Cumming didn't mention that, several days prior, Alice Hamilton and Paul Reznikoff had also asked for a government hearing on tetraethyl lead. Hamilton particularly noted "the desirability of having an investigation made by a public body which will be beyond suspicion." Hamilton thought that such a hearing and investigation would demonstrate all the ways Ethyl gasoline endangered the public.

Cumming lingered in his office after the men left. It was quiet in the building and quiet in Washington in the remaining hours before Christmas. It rained all day, and as evening set in, snow began to fall.

Before Cumming left to join his family for dinner, he looked at his calendar and made up his mind. Five months from then, in the third week of May 1925, at the headquarters of the Public Health Service, all parties on all sides of the Ethyl gasoline controversy would assemble in one room, air their differences, and come, at long last, to some semblance of a resolution.

# Part Four
# THE
# THUNDERING
# HERD

# 13

## Dayton, Ohio — 1925

The winter of 1925 was unusually cold in Ohio. All across
the Midwest, temperatures in January dropped below
zero and stayed there for weeks, freezing towns and railroad
tracks under feet of snow. Columbus had it the worst. A small
town on the east side of the city dropped to negative 40 degrees, among the coldest temperatures ever recorded in the
state. On the west side in Dayton, schools closed, factories
shut down, and businesses stayed shut for the first two weeks
of the year.

Things were also frozen at Ethyl Corp. The intense media
focus had largely passed with the arrival of the new year, and
as it did, Ethyl was surveying the damage to its product and
reputation. The board remained upset at such a public
and avoidable crisis. Alfred P. Sloan, the CEO of General Motors, had spoken with Irénée du Pont, the CEO of DuPont, and
Walter Teagle, the president of Standard Oil of New Jersey.
They represented the three principals of the Ethyl board of
trustees. The three men agreed that the blame for the mess

rested with Kettering and Midgley, whom they called "the boys."

Technically, the boys answered to the full board. But if they had a single boss, it was Sloan, who had supported their initial research and promoted their findings in an effort to boost General Motors. Since then, however, Sloan's esteem for Kettering and Midgley had diminished. In 1920, Sloan greeted the boys in their Dayton lab with smiles and back-slaps. But by 1925, Sloan had come to believe that they had risen beyond their competence.

Sloan knew they were talented research men. Both had proved successful at invention and discovery: Kettering with his automobile self-starter and Midgley with his laborious tri-als for an anti-knock fuel. Yet somewhere along the way they had morphed into businessmen. Kettering was taking steam-ships to Europe to discuss partnerships, and Midgley was holding defensive New York press conferences. They were now marketers and public relations hacks, and their limited expertise in these matters was evident every day that the news of men who died remained in the newspapers. "We felt that it was a great mistake to leave the management of [Ethyl Corporation] so largely in the hands of Midgley who is en-tirely inexperienced in organizational matters," Sloan wrote to du Pont that winter.

Midgley and Kettering could probably sense their goose was cooked, but it would take several more months for Sloan to work up the nerve to give them the boot. Kettering and Midgley were not outright fired, but they were replaced as president and vice president of Ethyl Corporation and sent back to their Dayton laboratory.

For the cause of demotion, the board could simply point to the numbers. Before the Bayway meltdown, the future of Ethyl seemed "bright" and "increasingly assured," accord-ing to an internal newsletter from October 1924. That month

Ethyl had reached its first monthly profit of around $500,000 and the fuel was on its way to broad market acceptance. But in the five months after the disaster, the monthly profits had dropped to $200,000 and then a $61,000 loss. These were small numbers compared to the earnings of Ethyl's multimillion-dollar parent companies, but none of the three horsemen atop Ethyl's board had gotten rich by backing losing stallions.

Kettering caught a lucky break, however. An investment he had made before the Bayway incident would soon yield dividends and buy him and Midgley extra time.

Several months earlier, in the summer of 1924, Kettering noticed something about the scientists who publicly criticized Ethyl. There was Alice Hamilton, the head of the field of industrial toxicology, and Yandell Henderson, the outspoken bulldog from Yale who warned at every opportunity of the deadly dangers of Ethyl gasoline. But largely silent on the matter was a young toxicologist at the University of Cincinnati named Robert Kehoe.

Kehoe was thirty. He had graduated from medical school at the University of Cincinnati and was eager to make his name in the growing field of toxicology. He had already published articles about toxic substances and public health. But perhaps on account of his youth, he, unlike Hamilton and Henderson, was wary of taking a stand or wading into public debates. This was probably why he avoided making controversial comments on lead, tetraethyl lead, or Ethyl gasoline. Kettering saw this as a plus. And it was an even bigger plus that Kehoe was an Ohio boy, like Midgley, which meant, ideally, that he wasn't corrupted by the snobby groupthink of northeastern universities.

Kettering thought Kehoe had the knowledge and credibility to flood the country with proof that Ethyl gasoline was safe. Or in other words that he could be useful in the fight for

Ethyl's future. He was local, just an hour's drive from Dayton, and could access Ethyl's office and labs. What was more, Kehoe had just gotten married, and he and his wife were trying for a baby, which made him especially open to extra money from consulting work.

Kettering contacted Kehoe in August 1924 and asked him to perform a few studies on the health effects of Ethyl gasoline. He said that Ethyl Corporation would pay for the work and reimburse any expenses, including Kehoe's personal salary at the University of Cincinnati. Kehoe did not hesitate. He signed a contract agreeing to perform two preliminary reviews of the existing science.

In October, Kehoe took his first look at the toxicity of tetraethyl lead. It was undoubtedly poisonous, he found. Based on a review of existing literature, he concluded that tetraethyl lead caused insomnia, vomiting, vertigo, and debilitating muscular weakness. This was not in dispute, which explains why Kehoe declared it so frankly in his first official paper on the matter. "It has been shown experimentally that tetra-ethyl lead is capable of producing illness and death of animals when inhaled as a vapor, when absorbed through the skin, and when administered orally or intravenously," he wrote. "Under any one of these conditions death may be brought about in the course of from three to six hours, though minimal lethal doses kill in from twenty-four to seventy-two hours, ordinarily."

To Kehoe, the question of Ethyl *gasoline*, however, was more complicated—or, rather, much simpler. There seemed to be no evidence published anywhere indicating that Ethyl gasoline produced any harmful effects to the broader public. Based on this limited review, Kehoe came to the unimaginative conclusion that the toxic danger of tetraethyl lead ended when it was mixed in small quantities with gasoline.

Kehoe wasn't outright corrupt; his findings and methodology were rooted in earnest and honest methods. But they were also lazy. Kehoe looked at the existing research about tetraethyl leaded gasoline, of which there was little, and concluded that the absence of evidence was sufficient proof that no broad danger existed.

Kehoe was staking out a new area of industrial toxicology, one that would become exceedingly attractive to America's polluting companies. Rather than a company proving that its invention was generally safe before introducing it to the public, Kehoe's subtle shrug suggested that a company could shift the burden to the public to prove the invention was dangerous. If the public couldn't prove it was, then why should a company stop making it?

Kehoe's work for Ethyl Corporation would later inspire a common corporate maneuver. A scientific theory in the 1980s named the "precautionary principle" would argue that new products should not be allowed to be produced and sold until sufficient research could prove their efficacy and safety. But the opposite principle—that a toxin is presumed innocent until proven dangerous—is now known as the Kehoe principle.

Kehoe proved himself extremely useful to Ethyl Corporation—not just for his selective research but also for the counterbalance he provided. Thanks to Kehoe, Ethyl could respond to the outraged comments of Alice Hamilton and Yandell Henderson in newspaper articles by pointing to Kehoe's more measured conclusions that were published, like all other research from white men from elite universities, in the *Journal of the American Medical Association*. No longer was there a unified group of scientists urging caution. Now there was *debate*, and within that debate a sense of doubt. How could Ethyl gasoline be unsafe if the nation's top scientists were torn over it?

Doubt was a valuable currency because it bought time. As long as scientists couldn't agree, there was no reason for Ethyl to shut down completely. And reason instead for it to continue its slow and quiet expansion.

A free press was one of the hallmarks of American democracy that made public debates possible. But the crowded media industry of the 1920s did not guarantee objectivity or even that the reported news was true. Papers that strived for accuracy could lose readership to sensational headlines and yellow journalism in other papers. Partisan newspapers made this worse by cherry-picking facts and perpetuating half-baked stories. Meanwhile, a lack of basic fact-checking or a formal standard of journalistic ethics led to endless errors that often went uncorrected.

The *New York Times* largely led the pack for covering big stories authoritatively. It was a centrist and moderate newspaper that leaned pro-business but had an overall reputation for impartial reporting. The *New York Post* skewed progressive. Both had a habit of producing sensationalist headlines, but none more than the *New York Daily News*, which laid the foundation for the city's tradition of tabloid reporting.

The *New York World*, however, was a holdover from an earlier era and staked its reputation on factual reporting rooted in social justice. The *World* was the creation of the media baron Joseph Pulitzer, who bought the failing newspaper in 1893 and transformed it into a circulation juggernaut. He slashed the newsstand price from two cents to one cent and focused on shocking news stories, crime reporting, and human-interest tales that appealed to a broader audience. He hired a team of investigative reporters to expose corruption and wrongdoing in a journalistic practice known as muckraking. Most notably, he used the editorial pages to express his own political opinions. A decade before most other papers, Pulitzer's *World* was advocating for labor rights, civil rights, and women's suffrage.

By 1925, Pulitzer was gone and the *World* had been passed to his three sons. The paper, which was now printed in eye-popping color, stayed edgy with deep investigations, including a lengthy 1921 exposé on the inner workings of the Ku Klux Klan. It also added a word puzzle with clues to words aligned horizontally and vertically. But the paper's soul hadn't changed. Its large circulation, combined with its historic reputation as a paper for the working class, gave it an impulse to pursue any story in which the common man was getting screwed.

That made Ethyl a perfect story for the *World*. And in 1925 the paper declared a public "crusade" against Ethyl and its "looney gas." Unlike the corporate-friendly *New York Times*, which generally accepted the rationales of Ethyl's top brass, the *World* pursued the company hard and skeptically into the winter and spring of that year. Rather than take the industry spokespeople at their word, the *World* far more frequently sought the viewpoints of university researchers, who usually spoke in scientific jargon and shared piles of past studies for the *World*'s reporters to decipher. The *World* was remarkable at the time, but its unique approach underscored how much easier it was for overworked journalists at other papers to simply accept the claims of major corporations and print them as fact.

Walter Lippmann was behind the *World*'s crusade, and his skepticism was the direct result of Hamilton's pleas a month earlier. She had written to him for sympathy and suggested what his powerful editorial page might advocate. "Given the shortcomings in the experiments and the data [in the Bureau of Mines study], I conclude it important that a new study begin at once," Hamilton wrote.

Hamilton had written to the right person. Lippmann saw journalism as a way to inform and educate the public, which was crucial for a healthy democracy. He was driven by an unflappable skepticism. He believed that anyone in power had strong incentives to stay in power and that those incentives usually ran against the interests of the people over whom they had influence. Lippmann used the large readership of the *World* to shame and embarrass officials not serving the public good. Sometimes this was exposing stupidity or ineptitude, such as when he wrote about Calvin Coolidge: "Nobody has ever worked harder at inactivity, with such force of character, with such unremitting attention to detail, with such conscientious devotion to the task." Other times it was simply pointing out—very publicly—when a company was too arrogant to respond to a set of allegations. The notion that a subject would look bad if a newspaper reported they "did not respond to a request for comment" was a creation of Walter Lippmann.

With Lippmann's influence, the *World* diverged from the rest of the New York papers. In February 1925, it reported that another worker, a man at a DuPont plant in New Jersey, had died from tetraethyl lead poisoning and the company planned no investigation or even an apology to the family. Several weeks later, it reported that a Harvard review of the Bureau of Mines study undermined the findings of the government inquiry. The paper quoted Cecil Drinker saying the bureau's experiments "did not show the substance safe for general use."

And then finally, a week after that, Alice Hamilton got exactly what she asked Lippmann for. There, above the fold of the opinion section of the April 12 edition of the *New York World*, was exactly the editorial she thought could add more public pressure to Ethyl and change the tide of leaded gasoline.

"The experiments conducted by the Bureau of Mines did

not use animals in sound health and were not sufficiently thorough," the paper wrote. Lippmann, who penned the editorial, concluded that prior investigations were "inadequate" and included "a great many inexact data." He ended the editorial with almost the exact words Hamilton had written to him. "A new investigation should begin at once."

Ethyl Corporation was the kind of American corporation Lippmann loved to pursue: the kind that seemed to be hiding something.

To have any shot at dislodging Ethyl gasoline from the American market, Hamilton and Lippmann felt they needed to ask a different set of questions. No longer was anyone examining whether tetraethyl lead was dangerous. That much was assured. Hamilton and Lippmann believed that if Ethyl gasoline was given the expert investigation Hamilton had long wanted, it, too, would be proven to have the damaging effects that she long believed it did.

The better question, she and Lippmann knew, was one of alternatives. If Hamilton or any of her like-minded peers could identify a *different* fuel that could eliminate engine knock without any lead, and ideally if it could be bought for the same price or even cheaper than tetraethyl lead, then they could beat Ethyl gasoline in the marketplace without all the work of a long investigation. Eroding its profit would render it obsolete.

Of course, there *was* an alternative. There were many, plain as day. As people had known for decades, the best anti-knock additive was ethyl alcohol distilled from grain crops, plant roots, or wood pulp. In 1920, while Midgley was still mixing test tubes in his lab, *Scientific American* called ethanol "the fuel of the future" and extolled the way that adding 20 to 30 percent ethanol to gasoline could completely eliminate engine knock. Ethanol still had supply issues in 1925, and there was the pesky problem that alcohol was illegal under Prohibition, but it did eliminate engine knock without poisonous tetraethyl lead.

There was also iron carbonyl, a gasoline additive that German chemists were testing that was cleaner and cheaper than tetraethyl lead. The Germans were so enthused about its ability to eliminate knock that they demonstrated it to Kettering during his 1924 trip to Europe, and even he was intrigued enough to send a telegram back to Alfred Sloan declaring iron carbonyl "so very interesting." (Much more interesting, however, was the large sum General Motors had already invested in making and marketing tetraethyl lead.)

And there were more solutions too. Gasoline mixtures with other chemicals like benzene, toluene, and xylene could reduce knock without the harmful effects of lead. There were acetone and manganese compounds. Midgley had tried versions of all these during his 1921 research quest. He dismissed them due to their various downsides: things like high freeze points, low boil points, or the fact that they'd leave gunky residue on engine parts. But those could be overcome with a little imaginative chemistry.

Hamilton talked about fuel alternatives with anyone who would listen in the winter and spring of 1925. She was invited to give more than a dozen speeches in Boston, New York, and Chicago to tell stories of her visit to the Soviet Union. At many of them she would eventually turn the discussion to U.S. matters. She spoke of an "injustice" unfolding in American business and that the public "wasn't given all of the facts." Her audiences sometimes nodded along, but at the end of one of her lectures a person could reasonably leave the auditorium wound up but unsure of what exactly they were supposed to do.

This was a familiar challenge for Hamilton. Her work was highly technical, and her ideas for how to solve problems were usually even *more* complicated. Her insistence on scientific precision made communicating with the general public almost impossible. Even just the name of her field, "industrial hygiene," was woefully clunky and meant little to the average person. (The field attempted a name change in 2020 to "occupational and environmental health and safety.")

The public was gripped by newspaper headlines about death by insanity and corporate conspiracy, but when it came time to advocate a regulatory or policy response, Hamilton usually lacked the words to persuade. She was an industrial hygienist talking about molecular compounds with *multisyllabic* Latin names. It was easy to say Ethyl gasoline was filled with toxic poison; her audiences fully believed that. But convincing people that they should use their power as customers to demand changes to a chemical that made driving more pleasurable was like asking them to demand that the New York Yankees eliminate home runs.

This confusion of consumers was a hallmark of 1920s business, and it worked especially well for America's chemical companies. Their work could not be easily understood or dismantled by anyone outside the field, often including senior government officers. Technicality of this kind clouded any attempts at regulation and tended to shift the role of public oversight to the companies themselves, as if to say, *We the government don't understand what you're doing, so we have to trust you're doing it right.* In this web of complexity, the tie often went to the corporation, especially when it came to bold new things that promised to improve modern life. Electric devices that could cure cancer, radioactive water that could alleviate headaches, and malt-based tonic that could reverse any ailment. When faced with a bold and exciting prospect, most people would not stop to ask questions.

Ethyl Corporation could anticipate its own public adoption as soon as it could move past the most recent controversy.

But it wasn't going to leave anything to chance. In the two months before the surgeon general's conference in Washington, Kettering and Midgley seemed to notice something that gave them great confidence. Even if they might face demotion to their former jobs, they at least saw a way to protect their prized product from being killed off.

Both men knew Ethyl gasoline had poisonous ingredients. They knew that tetraethyl lead had killed a dozen men and possibly more that hadn't been reported. But they also realized that in the four years since Midgley first invented the fuel, they had successfully demonstrated that none of the other fuel additives stood a chance at competing with the market reach and wide accessibility of Ethyl. And as long as no other fuel was coming to overtake Ethyl gasoline, they could muscle Ethyl through and collect its lucrative royalties.

While Hamilton and Lippmann schemed about how to dislodge a potent poison from the American marketplace, Kettering and Midgley were writing a playbook of their own.

At the core of their plan was doubt, denial, and delay. So long as they could leverage doubt to create a defensible environment of uncertainty, they could stall critical questions. When they *had* to answer critical questions, they could point to scientists like Kehoe for friendly findings. When faced with undeniable bad news or inevitable accidents, they could act contrite, without apologizing, and point to other companies and products that were even more dangerous, all the while paying off victims and their families with tiny fractions of their earnings. And when pesky doctors and newspaper editors accused them of dangerous and negligent behavior, they could discredit them by calling them naive, surfacing flaws in their past work, or suggesting they had questionable motives of their own.

In February 1925, Midgley wrote to Kettering that they must take "an uncompromising position" that any criticism of Ethyl was nothing but unpatriotic propaganda. Kettering responded that he fully agreed.

There was one final requirement to ensure the strategy would work: They would have to do it all with a smile. No one could accuse them of having evil motives if it seemed like they were happily and earnestly working toward a better future for the American consumer.

The upcoming conference called by the surgeon general would be the perfect opportunity to put these principles in action. And once it was proven effective, it would become the authoritative playbook for large companies that produced pollution or poison for a century to come.

14

*The surgeon general Hugh Cumming in his Washington office*

## Chicago, Illinois — 1925

Hugh Cumming had a mess on his hands. Almost every day, his secretary brought him piles of correspondence about the Ethyl matter. Cumming announced in January that the conference bringing together all sides of the dispute to duke it out would take place in May. In the five-month stretch leading up to the meeting, the expected attendees seemed to be jockeying for the upper hand.

A year prior, the debate over Ethyl gasoline was narrow. It was mostly a fight between industry and academic scientists, for or against. But the massive press following the Bayway disaster had tipped off millions more people to the controversy and driven many of them to share their opinions with the government.

On Cumming's desk were letters of outrage, letters of support, and letters wondering what all the fuss was about. Ethyl Corporation had gotten into the habit of sending Cumming a copy of all supportive correspondence it received during its crisis. So had Edsall and Drinker at Harvard, who denounced the Bureau of Mines' flimsy investigation and its faux exoneration of Ethyl. Cumming didn't subscribe to the *New York World*. He didn't need to. Almost every article the paper published about tetraethyl lead made its way to him in a letter from a person who was either furious about the poison or furious about the unfair criticism of it.

Hamilton visited Cumming twice that spring, taking detours from her constant travels to plead with Cumming that the upcoming conference needed to be fair and to stress that, without a disciplined format, Midgley and Kettering would likely steamroll the proceedings. "I observe it to be imperative that an evenness of time and discussion be given to all sides of this important matter," she wrote to him after one of her visits.

Meanwhile, Henderson at Yale had made destroying Ethyl gasoline an almost religious crusade. He gave fiery speeches about the fuel being nothing less than a slow-moving Armageddon that was a scarlet letter on America's national character. "This is probably the greatest single question in the field of public health that has ever faced the American public," Henderson said in April 1925. "It is the question whether scientific experts are to be consulted, and the action of the government guided by their advice; or whether, on the contrary, commercial interests are to be allowed to subordinate every other consideration to that of profit." The *New York Times* covered Henderson's speech, in which he labeled the effects of Ethyl gas "worse than tuberculosis." Three people clipped the article and sent it to Cumming.

Cumming had a broad philosophy about the nation's health. He believed that the primary goal of public health should be the prevention of disease rather than the treatment of patients who were already sick. Cumming thought that sickness and disease were driven by the social and economic factors transforming America. With this view, he would be the first public official to warn of the dangers of smoking and to speak out against the spread of contagious diseases on airplanes. His staff eliminated the curious custom of a "common drinking cup" at public water fountains. He saw it as the government's job to ensure that the average American wouldn't get needlessly sick and that companies were generally playing by the rules.

But Cumming's actual powers were vague. He couldn't pass legislation or even enforce laws that existed. The office was almost entirely rooted in leveraging the gravitas of his title to draw attention to things that could be dealt with by more senior officials. In this capacity he was often called to appear at events to make grand pronouncements that reporters would scribble on their notepads. In the winter and spring of 1925, while his office was receiving the flood of Ethyl-related correspondence, Cumming crossed the country twice. He scratched his head with scientists about infected oysters in the Chesapeake Bay and then boarded a train to California to stroke his chin with farmers about the millions of rats destroying the state's farms.

Owing to his limited authority, he knew that the meeting in Washington would not be a legal one with criminal or even civil implications. The Public Health Service could not regulate interstate commerce of a substance or enforce a national ban. Both would fall to Congress. But it was his duty, as he saw it, to "investigate such questions and to inform the public as to the result of its investigations," he would later say. If he did this successfully, the agency's findings would help the federal government, as well as state and municipal officials, to decide whether to allow Ethyl gasoline.

One wrinkle in this logic, however, was that adjudicating a national economic and health controversy would normally fall to Cumming's boss, Treasury Secretary Andrew Mellon, who oversaw the Public Health Service. But Mellon wasn't available, nor was *his* boss, President Calvin Coolidge. Months earlier, Mellon sensed it could be awkward if he got involved in the matter, since the company he founded and that made him rich, the Gulf Oil Company, had an exclusive contract to distribute Ethyl gasoline. Coolidge, for his part, had just won reelection in the fall of 1924 and had his mind elsewhere. Coolidge also hated the idea of the government doing anything to make it harder for businesses to operate.

Down the street at the Department of Commerce, Secretary Herbert Hoover also could have outranked Cumming, since Ethyl gasoline was a matter of public interest that involved the nation's commerce. But he told the *New York Times* that he "had not been asked to take an investigation into the poisonous or non-poisonous qualities of tetraethyl and did not contemplate entering into the present controversy." Hoover's deflection was notable, especially considering his agency was working on a report on alternative anti-knock fuels being used in Europe. But it made sense to anyone who knew that Hoover had met with Kettering several months prior, after which he promptly fell quiet.

That left Cumming to handle the Ethyl debacle almost entirely alone.

If Cumming felt sorry for himself, he had no one to blame.

His current bind was of his own making, and his lack of decisive action had allowed the matter to grow into such an unseemly mess. For years he had played both sides. He carried on chummy friendships with Alice Hamilton and Yandell Henderson and sought their opinion on broad matters of health and medicine. And he extended the same collegiality to Kettering and Midgley. He took their every call, answered their every letter, and when one or both arrived in his office, they'd begin with guffaws and backslaps before getting down to business. If he was biased in the matter, neither side could tell. But such impartiality seemed to give him heartburn about the showdown over which he would soon preside.

He didn't know how the meeting would pan out, but he knew enough to conclude that, no matter what he did, at least one side, and possibly both, would leave angry.

The Ethyl board finally worked up the nerve to demote Midgley and Kettering on April 21, 1925. By any reasonable standard, the boys had a disastrous run as business leaders. It wasn't their fault that their success as scientists had vaulted them to the upper levels of corporate governance. But it was their fault that their prized product, Ethyl gasoline, had become such a sweeping fiasco.

That spring, three more men died at the DuPont Ethyl plant in Deepwater, New Jersey. That brought the total to thirteen men killed while making a product that Midgley kept claiming could be made perfectly safely. A month later, as accidents continued to happen at the Deepwater plant, DuPont shut down the plant and was prepared to wash its hands completely of Ethyl gasoline, including giving up DuPont's seat on the board of Ethyl Corporation. But Alfred Sloan, the executive atop General Motors, convinced Pierre du Pont to hold off until the Public Health Service conference a month later.

There was no point in doing anything until then, Sloan said, holding out the slightest hope that Ethyl gasoline could somehow be vindicated or rehabilitated. Du Pont agreed. But his company was already making plans to move past Ethyl. Several weeks earlier, du Pont had signed a licensing deal with a German company to make and import iron carbonyl.

Kettering was "violently opposed" to being fired, as one biographer put it. He likely saw the move not as a demotion in pay and title but as an embarrassing censure that would mar his career. For two decades Kettering's stock had risen sharply due to his relentless streak of new inventions and lucrative patents. Now, at forty-eight, Kettering was the American archetype of a celebrated man who had grown too confident in himself. He thought that getting demoted would forever be an inflection point in his biography, one from which he might not recover.

Midgley seemed to care little about a change in title. Midgley was a scientist who grew up wanting to emulate the inventive spirit of Thomas Edison, and, at thirty-six, he felt that he had succeeded. Holding press conferences and arguing with reporters were not his strengths. Besides, it was getting harder to hide the effects of working with pure tetraethyl lead for more than five years. His hands hurt and his joints ached.

Kettering and Midgley asked that their demotions be carried out quietly to spare them the embarrassment. The board agreed, hoping to also spare Ethyl the bad press that would further imperil its reputation. Not a word was said publicly about Midgley and Kettering losing their top jobs. The only way anyone outside the company would find out was several months later when the letterhead of Ethyl Corporation changed one day to list Earle Webb, a former lawyer for General Motors, as the new president of Ethyl.

At the same meeting, the Ethyl board made another big decision. Less than a month before the surgeon general's conference, Sloan suggested that Ethyl ought to suspend all pro-duction and sale of gasoline until after the surgeon general held his conference and conducted an inquiry. To anyone watching from the outside, this was a surprising show of cor-porate prudence from a company that had knowingly man-ufactured and marketed an undeniable poison for almost four years. But it was also a self-serving act of reputational management. Halting production at a time when people were nervous about Ethyl gasoline came at minimal cost. But going into the Public Health Service conference with opera-tions already shut down conveyed a sense of seriousness and an openness to the facts. No one could accuse the company of profiting off death so long as, for the moment, it wasn't profiting.

None of this stopped Midgley from continuing to defend Ethyl gasoline. It was his creation, after all, and he felt protec-tive of it. Midgley had a son and a daughter, but he joked to friends that he considered Ethyl his third child. No amount of bad press and no amount of lead palsy in his hands could make him walk away from an invention that he deeply be-lieved was the answer to America's car-driving future and, rather conveniently, a product that had already proven ex-tremely lucrative to him and his employer. In a speech to the American Chemical Society in April of that year, Midg-ley defended the fuel as he usually did. But he went a step further and claimed that tetraethyl lead was the only anti-knock substance known to exist. "So far as science knows at the present time, tetraethyl lead is the only material avail-able which can bring about [an end to engine knock] . . . its abandonment cannot be justified," he said. His statement was not only demonstrably false but Midgley knew it, thus making it a "bald-faced lie," according to one historian. He knew about the Department of Commerce report in progress about anti-knock alternatives and he had received a memo about DuPont's investment in iron carbonyl from Germany. His statement was a preview of what he would argue at the conference in Washington and the lengths he would go to prevent any other anti-knock alternative from getting a public airing.

Such blatant dishonesty marked a new chapter for Midgley. For five years he had fiercely defended tetraethyl lead because he appeared to believe it could be made safely and that Ethyl gasoline itself was harmless. But in the spring of 1925, as counterevidence was piling up in the form of dead workers, he changed his strategy to ram his invention through any public opposition, no longer caring if he lied. He dismissed Yandell Henderson as financially motivated, even though his salary at Yale was unaffected by the controversy, and he rejected Alice Hamilton as naive about chemistry despite the thousands of pages she authored on chemical toxins.

Under so much public scrutiny, Midgley might have chosen to get off the treadmill, take a breath, and resolve that he had created a deadly mess.

But he and Kettering had invested too much. Rather than approach the matter with the earlier sense of scientific curiosity that had driven his work as a younger man, Midgley now saw his prized invention as under existential attack. And at the upcoming conference, he would use any trick and make any claim necessary to keep Ethyl, and himself, moving forward.

While Midgley and Kettering were preparing their defense in Dayton, Alice Hamilton sat in a dark movie theater on the other side of Ohio.

Hamilton was on her way from Boston to Chicago and had stopped in Youngstown at the request of General Electric. GE, as it was known, was a booming corporation working to transform America with electric motors, generators, lights, and home appliances. In Youngstown, it had a lamp factory that made tungsten filaments for light bulbs, and it wanted Hamilton to inspect the factory's health conditions.

Hamilton could see that Youngstown embodied the ruinous effect that ballooning numbers of cars were having on American towns and cities. Youngstown was a charming colonial settlement before it boomed after the discovery of coal. It grew even more after the construction of the Erie Canal and the laying of railroad tracks that passed through the town. In 1900 it was a bona fide city, but by 1925 any casual visitor could tell that automobiles were clogging the streets and filling the air with black smoke. "Youngstown is hopelessly ugly and there are no traffic regulations and I hated to be killed in such a squalid spot so as a last resort I went into a movie theater," she wrote to her aunt during her short stay.

Hamilton watched black-and-white images flutter on the screen in a silent film called *The Thundering Herd*. It was a story of American Indians in 1876 on the prairie and a band of ruthless outlaws who robbed them and stole their land. The film's sympathies were with the white men, the robbers, who embodied the future with their advanced weapons and their noble goals to build outposts and towns. But Hamilton saw the subplot. "It is made perfectly clear that the Indian fought because the whites slaughtered the buffaloes and so took away his only meat," she observed. There were graphic images of Native Americans dying and children crying for food—the poor and marginalized, overtaken by those with power. Sitting in a silent theater in the middle of the day, Hamilton recognized the same themes she had come to see frequently in her decades of fighting for the unseen. It was both a validation of her work and a depressing confirmation that the injustices of the world had existed long before her lifetime and would certainly endure afterward as well.

Hamilton felt prepared for the surgeon general's conference, where she would go after brief stays in Chicago and
New York. She packed almost an entire suitcase of scientific
publications demonstrating the damaging effects of lead gas
and dust on every organ. By luck, the *Journal of the American
Medical Association* agreed to publish Hamilton's most powerful public rebuke of Ethyl gasoline on May 16, just four days
before the conference. In it, she concluded, "Because of the
enormous and increasing use of automobiles, the question of
the danger to industrial workers and to the public which is
involved in the production and handling of tetra-ethyl lead
and the use of Ethyl gasoline is of the highest importance and
calls for a study which will be beyond criticism."

If challenged on this point, she could point to dozens of
other studies, many performed herself or by other Harvard
colleagues, to show that nothing about tetraethyl lead was
safe or ever could be. She was prepared to be attacked and
ridiculed. It wouldn't be the first time. In her career, now span-
ning more than thirty years, she had grown familiar with the
aggressive way companies tended to react when their executives—or their profits—were challenged or threatened.

She was grateful for one thing, though. Despite her cer-
tainty that tetraethyl lead was a public menace, as Henderson
had called it, the toxic substance had at least one redeem-
ing quality: It made people look. Every day in America, more
than 2,000 workers were injured in industrial accidents. Two
dozen died *every day* on factory floors. Newspaper reporters
hardly cared. Factory injuries and deaths were both woefully
abundant and painfully boring.

But now, when she was fifty-six, a substance had come
along that was so toxic and shocking that Hamilton's work
was finally getting the mainstream press coverage she always
thought it deserved. She mused to a friend that there was
something that appealed to the public's imagination about in-

haling a gas that would lead to insanity and a violent death—
and that the news reports might be a powerful weapon to
dislodge it. "Publicity is a wonderful thing," she wrote in the
spring of 1925 regarding tetraethyl lead. "It may be the pebble
with which David will kill Goliath."

She took no pleasure, however, in having to go to Washington to attend a conference on a question that to her was
not in dispute. There were not two sides to a poison being
poisonous.

She took even less pleasure in knowing that in barely two
weeks she'd come face-to-face with the two men who embodied all that she saw wrong with American capitalism.
And when she did, she'd tell Thomas Midgley and Charles
Kettering exactly what she thought of them.

# 15

## Washington, D.C. — 1925

On May 20, 1925, between nine and nine thirty in the morning, every expert on every side of the Ethyl matter walked through the front doors of the Butler Building at the corner of 3rd and B Streets in southeast Washington.

There was Kettering and Midgley, along with several top officials from General Motors, Standard Oil, and DuPont. Robert Kehoe was there too. Several organizations had sent representatives, including the American Petroleum Institute, the American Institute of Chemical Engineers, and the American Federation of Labor. The heads of at least six government offices were also in attendance, bringing together the Bureau of Mines, the National Bureau of Standards, the Bureau of Chemistry, and the Chemical Warfare Service.

Alice Hamilton arrived early—she always liked to be early—and found her way to the small scrum of her Harvard colleagues. The dean of Harvard Medical School, David Edsall, was there, along with Cecil Drinker and Joseph Aub, all of whom had studied or critiqued Ethyl. None of the Harvard men had been as vocal as Yandell Henderson of Yale, who came prepared to hold nothing back in his remarks. Henderson brought a young physiologist from Yale named Howard W. Haggard, who would help Henderson quickly recall any number of the thousands of pages of documents that might be referenced on any matter of lead poisoning or the inhalation of noxious fumes.

The odd assembly of people was taking place in an odd place. The Public Health Service had its headquarters in an old mansion, the house of the former Massachusetts congressman Benjamin Franklin Butler, who built his bespoke mansion across the street from the U.S. Capitol in 1871 on land he bought at auction. When the thirty-seven-room house turned out to be too large for one man, he sold it to the U.S. Treasury in 1891 for $275,000. Butler's former bedroom eventually became Cumming's office.

The building had a sense of grandeur of an earlier time. While new government offices were being built with Greek columns and art deco flourishes, the Butler Building, as it was named, still had parquet floors, hardwood moldings, and wainscot paneling in every room. Enormous frescoes, some thirty feet across, featured images of America's colonial period and its victory in the war of 1812. The building still had the feel of a house, not a government office, and so the work of Cumming's staff was carried out in bedrooms and closets. The only room big enough to hold a meeting of one hundred people was the ballroom, and so that's where all the attendees were ushered as the clock crept toward ten.

The day before, as Cumming prepared for the meeting, he wanted to place the chairs in a circle to give the gathering the feel of a roundtable committee discussion. This was classic Cumming, finding ways to include everyone as equals no matter their viewpoint, qualifications, or motives. But an aide told him that the size of the group—eighty-seven participants, with dozens of aides—made this impossible. He relented to placing the chairs in rows all facing a podium, thus giving the room a power center to be occupied by whoever was speaking.

"I will ask the meeting to come to order!" Cumming announced a few minutes after ten. People began to take their chairs. Cumming looked out over the room.

"It is perhaps advisable to say a word in the beginning as to the reason for inviting you to a conference and the object which we hope to gain as a result of your advice and counsel."

He looked down at his notes, likely a little nervous.

"The Public Health Service has long been interested in lead poisoning," he went on. "We were therefore interested two years ago when we learned that one of the large corporations in this country in its endeavor to increase the efficiency of fuel used in internal combustion engines was experimenting with certain lead compounds to be used in commercial gasoline."

Cumming acknowledged how contentious the issue had gotten. He mentioned how many letters he had received and how he was urged to take seriously the potential dangers in the general use of tetraethyl lead. With so many conflicting opinions aired in the media, he thought a conference should consider the matter, and, thus, here everyone was.

He concluded his remarks by noting that this was not a "legal hearing." Neither he nor his agency had any lawful authority to prohibit the sale of any substance, only to research and inform the public of possible threats. But it was his duty to help get to the bottom of the controversy and decide once and for all whether Ethyl gasoline was sufficiently safe or decidedly dangerous, and he hoped the group would work together toward that end.

"I am quite sure you are all going to help me and help each other arrive at a solution to this problem," he said.

To a casual observer, it appeared as though Cumming had opened the gathering with a spirit of impartiality—that all ideas would be heard and the best ones would win the day. But there were subtle biases already baked into the meeting before anyone else had even spoken.

For one, the Ethyl officials had the benefit of knowing extensively about Ethyl gasoline, both how it was developed and how it behaved when ignited in an engine. The detractors from Harvard, Yale, and other universities knew only what had been previously published about it, thus making the proceedings more of an exercise in call-and-response than an equal exchange of views.

There was also a temporal skew: Everyone defending Ethyl relied on a body of evidence from the past. They had the benefit of seeing what Ethyl was capable of, how it could go wrong, and how it could improve. Hamilton and her side, however, had come mostly to caution about Ethyl's effects in the future. There was no data about the future, only well-informed conjecture based on peripheral studies. It was impossible to prove that widespread danger to the average person on the street was imminent, since the widespread danger hadn't happened yet.

And there was something else. Cumming invited Kettering to be the day's first speaker, a privilege generally afforded the prosecution in a trial. This was not a formal trial. But it was an informal one on a substance under suspicion. And as Kettering walked to the podium, it signaled to the attendees that Ethyl and its representatives were equal partners in the matter and not, in fact, defendants needing to prove the innocence of their product.

Cumming prepared to introduce Kettering, a colleague who in the previous months had been so much in contact that the two men were almost friends. "I have thought it would be interesting, at least to some of us, if we were to be-

gin this conference by asking someone to give us the history
of the development of this tetraethyl lead in its relations to
gas engines," Cumming said. "Mr. Kettering will give us the
historical account."

Kettering walked to the podium. He was prepared to give
a short history of how Ethyl gas was developed as a foundation for a discussion about its safety.

"Mr. Chairman, ladies and gentlemen," Kettering bellowed.
He introduced himself as the president of Ethyl Gasoline Corporation, even though he had been demoted from the position a month before.

"About 1914," he began, "we undertook to determine what
were the essential factors in an internal combustion engine
which prevented us from getting more economy . . . We [were]
getting about 5 percent of the energy out of gasoline; 95 percent of it is thrown away. It is possible, however, to push up on
full load so that we obtain up to a maximum of 30 percent."

For the next twenty minutes Kettering explained how internal combustion engines worked. He took a long detour
into the chemical processes of iodine and then gave a selective description of the miraculous properties of lead, and tetraethyl lead specifically.

He tailored his message to convey the urgency of the fuel
situation in America, underscoring that tetraethyl lead was
a miracle substance—the *only* miracle substance—that could
rescue the country from an imminent economic collapse if
known supplies of gasoline were soon exhausted.

"We will use 12 billion gallons of gasoline this year and
15 billion next year," he said. To meet the growing need, he
continued, the only successful strategy was to make engines
bigger or fuel more efficient. And Ethyl gasoline was the only
fuel that proved it could be done on a large scale.

As Cumming had asked, Kettering then described the
nearly three-year process to find an anti-knock substance.

After all those experiments, he explained, tetraethyl lead was shown to be the sole substance that could cleanly and efficiently extend the mileage of gasoline.

When he was finished speaking, Cumming stepped back to the podium to pose the primary question of the day: We would like to have an answer, Cumming said, as to whether tetraethyl lead gasoline can be "manufactured safely" and distributed without incident.

Midgley took this as his cue. He assembled his notes and rose to speak.

Yes, it can, Midgley said. DuPont had a refined method whereby tetraethyl lead would be shipped in steel drums to regional filling stations where technicians could mix it with gasoline before selling it to motorists. This had occurred safely with almost all batches of Ethyl gasoline sold in the past two years. The incident at Bayway was not a reflection on the fuel but on the men who were poorly trained to work with it.

"Tetraethyl lead is not so much a dangerous poison as it is a treacherous one," Midgley said. "It becomes dangerous due to carelessness of the men in handling it. But by enforcing proper discipline, with mechanical devices to perform certain operations so that the men did not have any excuse for contact with the material, the record has become quite good." His implication was that the additional deaths since the Bayway incident were just more unfortunate souls who weren't following protocol.

The pace of the morning was starting to quicken. As though they had rehearsed it, Midgley's statements ended with a smooth segue to Frank Howard of Standard Oil of New Jersey, whose role, in several remarks throughout the day, was to make a broad appeal for the continued use of tetraethyl lead as a matter of social progress.

Howard pointed out that to the scientists like Hamilton

and Henderson the question of the conference was simple: Was this a public health hazard? Even if everyone could agree that it was to some degree, the question for the fuel industry—and for America broadly—was more complicated. "We are engaged in the General Motors Corporation in the manufacture of automobiles, and in the Standard Oil Company in the manufacture and refining of oil," Howard said. "On these things our present industrial civilization is supposed to depend. . . .

"What is our duty under the circumstances?" Howard continued. "Should we throw this thing aside? Should we say, No we will not use it [despite it being] a certain means of saving petroleum? Because some animals die and some do not die in some experiments, shall we give this thing up entirely? We can not justify ourselves in our consciences if we abandon the thing . . . merely because of our fears."

Howard's remarks were persuasive. By reframing the health question as a matter of human advancement, he slyly put Ethyl's detractors in the awkward position of opposing the destiny of humanity. Twice in his remarks Howard referred to tetraethyl lead as a "gift of God." (Someone would later point out that it wasn't such a godly gift to the men it killed.)

It went on like this for more than an hour. The "discussion" Cumming had promised was so far an industry advertisement with Ethyl's parade of executives and others paid to vouch for the product. Several more men spoke, including Robert Kehoe, the Ethyl-funded doctor from the University of Cincinnati, and Frank Morton, the medical director of Standard Oil of Indiana. They discussed studies they had performed and workers they had examined. Their message was unified and coordinated, all underscoring that past problems with Ethyl had been addressed and the data showed no harm would come from Ethyl in the future.

As an act of corporate public relations, Ethyl and its parent companies could not have asked for a better opening to a government inquiry into their operation that would be heavily covered in the media. After another hour, it was clear to anyone paying attention that all the remarks so far were about Midgley's ingenious discovery and Ethyl's selfless contribution to the American motorist. Barely a dozen words were spoken about factory accidents and dead workers. And no words at all about the effect of a well-documented poison on the public at large.

Alice Hamilton sat quietly and listened to these remarks. She was not prone to confrontation or breaks in decorum. If she seethed, as her letters later suggest she did during the conference at "the minimizing of what seemed painfully obvious," she kept it to herself.

Her colleagues, however, did not repress their frustration. By one o'clock, after three hours of Ethyl's one-sided corporate narrative, Cumming dismissed the meeting for a lunch break.

David Edsall and Yandell Henderson were infuriated that the first half of the day had been devoted to industry blather backed up by industry doctors toeing the industry line. A reporter for the *New York Times* stood nearby as the men talked and quoted Edsall in the paper the next day.

"It has been clearly shown that there is a hazard to the public in the use of Ethyl gasoline," he was overheard saying. He explained that the logical problem was that "we are not able to judge how serious that hazard is." The only way to reach a definitive conclusion was not through slapdash experiments with a few animals in cages but with wide-reaching trials extending years, maybe decades.

The other scientists agreed. It was obvious to all of them that the morning session was nothing but corporate maneuvering with tacit cooperation from a business-friendly government. As a result, the burden of proof seemed to be on the scientists to prove something that they lacked the time and resources to prove. And since they couldn't prove it, there was no proof.

And there was something else that bothered Edsall, Henderson, and Hamilton. All the industry men who spoke seemed to be in no hurry. They were all on the company dime, paid their company salary to defend their company product. But the scientists were on personal time. They had taken leave from their duties and their students. They had paid their own train fare to Washington and their own overnight accommodations. There were rumors the conference would go three days, maybe four. Why should they have to incur so much personal expense to sit at a conference that, so far, seemed to have little use for their insights?

The group agreed that after lunch Henderson would speak first. And that when he did, he should be forceful and blunt.

At two o'clock, Henderson took the podium like a preacher, with more to say than he had time to say it.

"I have been assigned to open the discussion, and I am going to be brief," Henderson said.

"We have in this room two diametrically opposed conceptions. The men engaged in industry, chemists, and engineers take it as a matter of course that a little thing like industrial poisoning should not be allowed to stand in the way of a great industrial advance. On the other hand, the sanitary experts take it as a matter of course that the first consideration is the health of the people."

Henderson explained that lead was the most common industrial poison known to humanity, at least until it was replaced by carbon monoxide with the advent of automobiles. Lead was everywhere: in paint, in furniture, in early electronics. It was impossible to know how much the average person accumulated, but there was more lead in more people than ever before in history. Henderson explained that he wasn't saying that Ethyl Corporation was responsible for lead poisoning in typesetters, printers, and housepainters. But if so much lead was already out in the world, any additional lead added to a person's body could pass a dangerous threshold, and that's what made Ethyl gasoline so dangerous.

Henderson paused to riffle through his papers and decide which direction to go. He brought letters he received from men who worked in Ethyl factories who detailed the lack of any precaution. He had personal notes he had jotted recalling his time overseeing a production facility making mustard gas in the Great War, in which no men died despite the extreme danger of the substance. He had evidence from an experiment showing how exhaust fumes were sometimes discharged *inside* the cabins of automobiles and how tetraethyl lead could poison an oblivious driver while he drove. And he also wanted to mention that even if plant workers were properly trained to handle Ethyl gasoline and motorists were properly warned to avoid it, no amount of precaution could prevent smart people from doing dumb things, such as how he had once observed his Yale colleagues on their boats sucking clogged fuel pipes and spitting out mouthfuls of gasoline.

The point was, there were dozens of ways Ethyl gasoline would be dangerous to almost everyone.

Henderson tried to assemble his disparate thoughts. "The main calculation which I want to contribute to this discussion is this: Doctor Haggard, my associate here, and I have published extensive work on automobile gases in the streets. We know the amounts of carbon monoxide in garages; we know that the figures run to one or two parts in the streets and eight to 10 or 12 parts in garages, and more than that in repair shops.

"Now, taking the data obtained by the Bureau of Mines and combining them with other data we have, it appears that in an automobile every kilo of gasoline burned involves 12 kilos of air. In the exhaust gas there would then be about one milligram of lead of each 0.8 liter of carbon monoxide produced, or 1.4 milligrams of lead per liter of carbon monoxide."

Henderson was speaking quickly, and it was difficult to follow his math. But he was driving toward his point.

"If, then, a man breathes four to five thousand liters of air in ten hours, and the air contains one part of carbon monoxide in ten thousand, as in the streets, he would inhale in the course of ten hours one-half milligram of lead. That, of course, sounds to a chemist like an exceedingly small figure. [But] to the man interested in industrial diseases the daily inhalation of half a milligram of lead is a serious matter. In a garage, where often there are ten parts of carbon monoxide, the worker would inhale daily two and a half milligrams of lead."

Two and half milligrams of lead exposure per day was the same exposure that was thought to befall the elite Romans who drove their empire into the ground. If that much lead was polluting the water supply of any major city, it would be immediately stopped. But lead in the air made it feel more diffuse, like it would be quickly blown away by a clean breeze.

Henderson spoke for fifteen minutes. As he was winding down, he called Hamilton up to elaborate on his remarks. No one could express this matter more fully and clearly than Hamilton, he said.

Hamilton gathered her papers and was beginning to stand up.

But before she could start walking, Kehoe took advantage of the pause and darted toward the podium. As a scientist, he had reasonable claim to speak during the afternoon session devoted to scientific discourse. But even if someone pointed out that he was an Ethyl man—bought and paid for to produce favorable results—and thus belonged in the morning session, his enthusiasm and energy could not be stalled in his rush to the front of the room.

"Mr. Chairman and gentlemen, there are one or two items of Dr. Henderson's talk that I should like to speak of," Kehoe said.

Hamilton sat back down.

Kehoe's maneuver was rude. But to respond to his lapse of decency with even polite firmness (*I believe it was my turn*) would bring the unwanted scorn of a woman speaking up. It was yet another instance of the perilous position of being a woman in the sciences, of being belittled and overlooked, with no recourse but to accept the insult with pursed lips.

Hamilton was so used to this treatment that it no longer bothered her. Only in the safe company of other women did she even remark on it. "I have so often felt myself pushed into obscurity and passed over that I have almost ceased to fuss over it," she wrote not long before the conference to her sister Margaret. "I wish I could say, never mind, in the end you will get your just dues. Unfortunately often one does not get one's just dues, they are grabbed and one cannot grab back."

At the front of the room, Kehoe breathlessly explained that the animals exposed to lead in the earlier experiments could not be compared to people. If anything, the animals had more lead exposure than the average person would, and even though they all died, their autopsies did not show debilitating lead poisoning. He also complained that the number of men who "thought" they were sickened by tetraethyl lead but were not—such as a report he heard of a man who owned a filling station that sold Ethyl gas and complained to his dentist about a toothache from an unrelated condition—reflected unfairly on the company.

With this, Kehoe had made his public evolution from a scientist to a corporate hack, laying the full weight of his medical reputation on the line for a company that he believed represented, as he called it, "progress." In this moment he was prepared to vouch not just for Ethyl Corp's data, as he was hired to do, but for its full corporate character.

"I am convinced from the association I have had with the [Ethyl Corporation] that their attitude is one of complete regard for facts," he said. "If it can be shown that an *actual* hazard exists in the handling of ethyl gasoline, that an *actual* hazard exists from exhaust gases from motors, that an *actual* danger to the public is had as a result of the treatment of the gasoline with lead, the distribution of gasoline with lead in it will be discontinued from that moment. Of that there is no question."

Kehoe's vow was absurd. Several minutes earlier, Henderson had cited numbers showing the actual harm associated with tetraethyl lead. Kettering's claim that no evidence existed was either a poor attempt at misdirection or perhaps a very successful one.

The conference paused for another break around three o'clock. And it was then that Hamilton had heard enough.

She walked over to the group of Ethyl men crowded together. Kettering was talking to Midgley and Kehoe.

In that moment, Hamilton's frustration boiled over. For three decades she had battled against slimy business bosses who prioritized their self-interest above all else. The smiles, the self-regard, the utter apathy that people had died on their watch because of their negligence. And, worst of all, the smug confidence that they would not only get away with it but that they would get richer in the process.

Hamilton stood in front of Kettering with fire in her eyes.

"You"—she pointed to him—"are nothing but a murderer."

Kettering was silent for a moment, perhaps bemused. He was not threatened by the short woman standing in front of him, no matter how smart or respected she was.

She was not finished. "There are thousands of things better than lead to put into gasoline."

Kettering looked at her for a moment, considering his retort.

"I will give you twice your salary if you will name just one such material," he said.

"Oh, I wouldn't work for *you*," she shot back.

Her response was weak, and when the Ethyl men returned from Washington, they would laugh about this exchange.

If Hamilton had a moment to think, she could have responded that iron carbonyl and ethyl alcohol were already proven to be superior to tetraethyl lead. She could have recited the fourteen names of the men who died handling Ethyl, or the obvious failures in the studies that investigated its safety. She could have even listed the precise way she believed Ethyl gasoline would poison millions of people across America, the lawsuits that would follow, and the stain it would bring on Kettering's and Midgley's legacies.

But it all happened so fast. And none of it came to her mind in the moment.

The men returned to their conversation and Hamilton walked away, a little embarrassed and steaming mad.

Hamilton finally took the podium around four thirty. The conference was in its waning hours, and the enthusiasm of the morning session had succumbed to the general malaise of the late afternoon.

Cumming noticed that almost the entire conference had passed and Hamilton had not yet spoken. If not for Cumming, who called her up while remarking that Hamilton "always has something worthwhile to say," she might not have been given or sought the chance to speak at all.

Hamilton had prepared a lengthy document of remarks.

But throughout the day she crossed out topics as many of the prior speakers made the same points she intended to. There was no point in being repetitive. All that seemed left was to underscore a few prior comments.

"I would like to emphasize one or two points that have been brought out," she said.

She began by remarking that lead was a slow and cumulative poison and that any effort to quickly detect and diagnose its presence was very difficult in a clinical setting and much harder among the wider population of people who might get sick and have no idea of the cause. She also declared that if she and her medical colleagues were even *partially* right and Ethyl gasoline was only *mildly* dangerous to the general public, then the damage would still be extraordinarily widespread.

There was no way to sidestep this issue, she said. "You may control conditions within a factory, but how are you going to control the whole country?"

Lead was always dangerous in every context, Hamilton said. She, one of the world's foremost experts on industrial lead poisoning, had never found a lead factory that was able to operate completely safely. Not one. She believed that for tetraethyl lead to be made and distributed without harm, it would require extreme and eternal vigilance at every point in the supply chain. And everyone in the room knew enough about the flighty American temperament to understand that the whole country couldn't remain eternally vigilant about anything.

Then she turned toward Kettering and Midgley:

"I would like to make a plea to the chemists to find something else."

This was the same point she made to the Ethyl men's faces. But now it was on the record in a government proceeding, essentially calling them lazy or liars, or both.

"I am utterly unwilling to believe that the only substance which can be used to take the knock out of a gasoline engine is tetraethyl lead."

It was possible to find an innovative solution to this problem, she said. She even had proof. She described how nine years earlier, during the war, England covered its fighting planes in a flame retardant known as tetrachloroethane dope, a poisonous substance that ended up killing twelve men. This was wartime, a moment when twelve deaths were a tolerable price for England's dominant airpower. But twelve deaths were too much for the U.K. government, Hamilton said, and it demanded that a substitute be found. Sure enough, a year later, tetrachloroethane dope was discontinued and replaced with a safer substance.

Hamilton and her side had made their points: Safer alternatives existed, the prior investigations into tetraethyl lead were weak, and Ethyl gas would be dangerous to everyone in America and possibly the world.

At the end of her remarks, Hamilton proposed a way out of this mess. Since Kettering, Midgley, and Ethyl Corp. could not be counted on to take any semblance of a higher road and change course, she declared that the solution was in Cumming's hands. Cumming and his colleagues in Congress needed to demand safer options to dangerous substances like Ethyl. And if they did, deep-pocketed companies like General Motors, Standard Oil, and DuPont could and would pursue new alternatives. She believed that any suggestion otherwise was false and rooted in self-serving apathy.

Several more men spoke after Hamilton, but there was little air left in the room. A spokesman for Ethyl again suggested that it would be irresponsible to set aside a "gift of God" just because some animals died in some experiments. Besides, he said, the conference had aired all sides, and maybe he was biased, he admitted, but it seemed that "most of the facts presented have been in favor of the use of the tetraethyl lead product."

Looking back in time, this last hour of the first and only major government conference about a toxic substance poised to take over America would be a master class in corporate bulldozing.

The strategy set by Ethyl's high-priced public relations—to repeat its position over and over, admit no weakness, and infuse doubt into all arguments from its opponents—was working. And with the complicity of a distracted government, powerful men could use their leverage almost limitlessly to preserve their power. In a decade that would become known for its dynamic economic boom and bust, there was nothing more American than the sense of infallibility of American companies.

As the clock moved past five o'clock, Cumming again took his place at the podium.

"Are there any further remarks?" he asked.

No one raised their hand. After six hours of speeches, pronouncements, and rebuttals, there appeared to be nothing more to say. And the silence suggested that not a single mind had been changed on the issue.

The meeting moved into its final phase: deciding on what to do next. Cumming suggested that if there was no consensus among the parties in the room, then there would be only one way to settle the debate: one final study, conducted under the supervision of the Public Health Service, which would report directly to him. The study would be the last word on the matter, and, depending on what it found, Cumming would recommend to Congress that it either outlaw Ethyl gasoline or allow it to proceed.

Francis D. Patterson, a doctor with the Pennsylvania

Department of Labor, submitted a resolution for Cumming to read and the group to vote on:

> *It is the sense of this conference that the Surgeon General of the United States Public Health Service appoint a committee of seven recognized authorities in clinical medicine, physiology, and industrial hygiene to present to him . . . a statement as to the health hazards involved in the retail distribution and general use of tetraethyl lead gasoline motor fluid, and that until such time distribution of this substance be discontinued.*

Patterson essentially called for one more study, free from Ethyl's corporate interference or money. Free from bias from anyone with any commercial interest in it. Complete objectivity, or however close the scientific method could get.

Under this plan, Cumming would choose committee members and give them almost limitless authority to use and scrutinize Ethyl gasoline. They could conduct experiments in almost any manufacturing facility, garage, or on any public street. And their results would be submitted directly to the surgeon general without any input from the industry.

There was just one wrinkle. Earle Webb, the newly appointed president of Ethyl Corporation, pointed out that the company had already paused the sale of Ethyl gasoline until the issue was resolved and that a lengthy study would cost the company to keep its production and machinery idle.

This was a reasonable complaint, but in the scope of a question of pollution that could have a decades-long effect on millions of people, it was a small one.

But Cumming agreed with Webb. Rather than give the study as much time as the committee needed, he proposed that the committee would have until January 1, 1926—seven months in all—to conduct a definitive investigation whose results would put this issue to rest.

Seven months would never be sufficient to conduct such an exhaustive series of experiments in so many locations on a matter of national and possibly international public health. But it was 5:45 p.m. in a room of tired people, and, besides, the academics felt like they had won a hard-fought concession by forcing an inquiry that Ethyl would not be able to influence or control. The resolution signaled that the final fight over Ethyl would be kicked forward seven months. Nothing more would be gained or lost by more talking that day. And so the meeting adjourned.

Hamilton walked out with Edsall. Behind her were Henderson and Drinker. They were all exhausted, but they didn't feel as though they had lost. There was still the chance, with an unbiased and publicly funded study, to prove the danger they all knew to exist. They wondered whom Cumming would appoint to the committee. They all agreed that, if asked, they would accept the job.

In the end, Cumming chose Edsall and another Harvard professor, Reid Hunt, along with professors from Johns Hopkins, Vanderbilt, Yale, and the University of Chicago. Despite Hamilton's leading expertise in lead and epidemiological experiments, the committee was all men.

Hamilton was used to getting slighted; any woman of the era would be. But she had other things to think about and other places to go. The next morning, she caught a train to Schenectady, New York. General Electric needed a survey of another one of its light bulb factories and had asked Hamilton to come take a look.

# 16

## Washington, D.C. — 1926

It took a year for Alice Hamilton to realize she had lost.

For a few weeks after the conference, it seemed like things were stacking up favorably, that victory might even be at hand. The sales of Ethyl gasoline were suspended, the government had agreed to investigate the matter, and the question had been wrested from the hands of biased industry officials and put in the hands of capable and independent health researchers.

Hamilton remarked a month after the conference that, unlike a decade prior, when the most she could do about a poison was meet with its maker and ask nicely for safety precautions, the surgeon general's committee was "a great progress" in how it could not be tampered with by the industry. She believed that the difference had come from the "blaze of publicity" from the *New York World* that exposed and sought to shame Ethyl Corporation.

But it became clear that these bright spots were a facade. And when she realized it, the feeling of defeat descended slowly. By the spring of 1926, it became untenable to not see the whole Ethyl controversy as mostly resolved, and not in a good way.

After the surgeon general's conference in May 1925, Cumming's committee of experts made several consequential decisions that almost instantly neutered their work. First, the committee members agreed to Cumming's request made at the urging of Ethyl president Earle Webb that they complete their work within six months, seven at most, to meet the tidy deadline of January 1, 1926. Webb wasn't nefarious in his pushing, just impatient to restart the idle Ethyl production line. Every day not producing was costing the company money.

Some of the committee's members, including Edsall and Hunt at Harvard, thought the study needed much longer, possibly two or three years, for a true national investigation. But there were also practical considerations, like the limited time that the scientists could reasonably take from their regular duties. Not to mention the money. It was a chore for Cumming to procure $50,000 from the bean counters at the Treasury Department. More seemed out of reach and likely impossible.

Meanwhile, the short runway for the study grew shorter as the days flew by. It took Cumming a month to formally appoint the committee and another for the members to design a study. Three more months passed while the committee secured access to several garages and plants in Dayton and Cincinnati, and more time to get the approvals and insurance for the investigators to work with such dangerous chemicals.

The investigation finally started in early October 1925, five months into the seven-month deadline. This left barely four weeks for the investigation and three more weeks to review results and write a report before everyone went home for Christmas.

Four weeks would have to be sufficient, and so the investigators reduced their original plan for an expansive study to one that would fit snugly in the span of a month. In that month, they studied 252 people, all of them men. They put those men into four main groups: 36 Dayton city employees whose job it was to drive with regular gasoline (the control group) versus 77 Dayton workers who drove daily with Ethyl gasoline (the test group). Another track would compare 21 garage workers and filling station attendants who worked with regular gasoline (control group) and 57 men who had similar jobs using Ethyl (test group). A true experiment would involve testing the men repeatedly to see how their blood levels changed over months, and even years, depending on their exposure, and then documenting any common symptoms or deaths that occurred among the groups. But there was no time for that. The researchers resolved to simply conduct onetime blood and fecal samples of each man and compare them to each other.

Despite these primitive research methods, the results still held striking revelations. Almost all the men in every group, even the control groups, had lead ash in their bodies (likely the result of lead exposure from paint, pipes, or other areas of life). But some of the men working with Ethyl had significantly more lead—more than 1 milligram—and a much higher breakdown of their red blood cells than those who did not work with Ethyl.

Such findings would normally inspire scientists to dig deeper and tinker with certain factors. What if some of the workers were exposed to substantially more Ethyl than they were during the test? How would the results change if rerun after six months of exposure, or one year? There were also strange inconsistencies, such as how the Ethyl gasoline samples used in the study (which had been acquired from a DuPont plant in Ohio) were found to contain only half to three-quarters of the usual amount of tetraethyl lead used in gasoline. This may have been a normal variation, or it could have been fraudulent tinkering to skew the results, but nobody had time to investigate.

By early December, with the clock ticking, the committee prepared its report. Its final conclusion, based on its limited work, was that Ethyl gasoline showed distinct signs of danger, but not enough to rationalize an outright ban. This framing was rooted in a lazy rationale of comparing it to other industrial chemicals. The world was filled with hundreds of compounds, many of them also lightly scrutinized. Ethyl did not appear any more or less dangerous than the worst ones, and thus it lacked sufficient grounds to be banned.

Put another way, a recommendation to ban Ethyl gasoline would bring the full weight of a growing and popular industry onto a mere seven scientists ill-equipped to fight back. And even if they escaped with their jobs and reputations intact, their suggestion to ban an industrial chemical would invite scrutiny on scores of other even more dangerous chemicals, like benzene, arsenic, and mercury. Was it worth it for the scientists to be forever blamed for disrupting American industry during its hottest ever decade?

Three days before Christmas 1925, the committee's morale was low. In a meeting in Washington, Edsall bluntly complained to the other members that they were submitting a "half-baked report."

Hunt agreed. In addition to producing such a weak study, both he and Edsall considered it irresponsible to make such a wimpy and vague conclusion about a supposedly important issue of American public health. The least they could do was disclose that more time and money would yield more robust results.

Charles Winslow, a committee member from Yale, agreed with them. Winslow feared the issue could reflect poorly on the scientists involved and potentially render them responsible in case of disastrous consequences to come. Fearing personal embarrassment and in an act of reputational self-defense, Winslow proposed including in the report that "a more extensive study was not possible in view of the limited time allowed to the committee."

Winslow was right to worry that his reputation and even his legacy were on the line. Historians would later dissect the committee's work and the political pressure it was under to understand why some of the nation's top scientists produced such an anemic investigation into what history would judge to be a sweeping matter of life and death. Many years later, a Johns Hopkins University science historian named Stuart Leslie blamed the breakdown on three clear factors: The analytical techniques were crude, the sample size was small, and the time frame was too short.

Predictably, a conclusion that no one on the committee felt strongly about ended up meaning little to the American public. A hysterical story that one year before commanded daily front-page headlines above the fold was now relegated to the inner pages. On Wednesday, January 20, 1926, the *New York Times* published the committee's findings in a small story on page 13. The paper reported that there were "no good grounds for prohibiting the use of Ethyl."

The matter was officially settled.

Hamilton didn't believe it was over. Surely a measly two-month study couldn't be the end of what had consumed years of her time. But the more she waited for another shoe to drop, or at least a public outcry for a real investigation like the kind she had done almost single-handedly in Chicago in 1911, the more it appeared no other horsemen were coming to rescue her—or rescue anyone. Even Walter Lippmann at the *New York World* admitted the battle was over when his paper published a story on January 21 declaring that the Ethyl inquiry had been resolved.

This realization was hard for Hamilton to swallow. For the

first few weeks after the committee's report, Hamilton held on to her belief—wishfully—that the whole affair had damaged Ethyl's reputation enough to weaken it in the market, possibly fatally. If non-Ethyl gasoline was for sale in an American town, why would a consumer take a chance on a fuel that had caused so much controversy? Besides, if a consumer in Boise or Durham cared to, he could read between the lines and notice that the Public Health Service's no-good-grounds-for-prohibition conclusion didn't exactly mean Ethyl was *safe*; it just meant it wasn't dangerous enough for the government to boot it from the marketplace.

Hamilton shared these thoughts in a letter to the Yale physiologist Howard Haggard. "There is a firm difference in declaring a substance safe for use and simply permissible, I do wonder if people will take note," she wrote. In Haggard's response, a copy of which he sent to the *Times*, he agreed that the panel's unenthusiastic half endorsement would poison Ethyl in the minds of consumers. "I doubt very much if the Standard Oil Company will manufacture it, even after the decision has been made in their favor," Haggard wrote. "There has been so much publicity in the matter that it will take a long time for people to regain their confidence in it."

But Hamilton's hope—and the hope of anyone else who knew how poisonous tetraethyl lead was—was quickly dashed. By early March 1926, Ethyl Corporation was finalizing an elaborate marketing campaign to reintroduce Ethyl to American drivers. General Motors boss Alfred Sloan mischaracterized the government report on Ethyl as a "clean bill of health." A year prior, Sloan was ready to abandon Ethyl completely. Now he was newly energized. The one-year pause in sales had cost Ethyl $3 million. But that was a small sum compared to the massive revenue Sloan could now foresee.

Sloan authorized nearly a million dollars in advertising. He ordered his company to buy billboards and signs on the sides of buildings, some of them adjacent to filling stations. Then he offered incentives to station managers to carry and push the fuel. By June, magazine ads and signs at Standard, Amoco, and Esso stations across America started to declare: "ETHYL IS BACK."

The advertising worked. On top of Ethyl's own marketing, mom-and-pop stations were taking out ads in their local newspapers (subsidized by Ethyl) urging customers to come check out Ethyl and see for themselves. They called Ethyl a "super fuel" with "no substitute" that promised "faster pickup and less gear shifting." Some stations offered the first tank free.

This wasn't deceptive marketing. It was true. When new drivers tried Ethyl for the first time, the promises checked out. It *did* stop the jerking during sudden acceleration and it *did* reduce shaking while climbing hills. More power, more speed, and no knocking. Whether consumers put it in a Ford Model T, a Chrysler Imperial, or a Buick Model 24, it made cars a pleasure to drive.

The moral gymnastics that Hamilton expected consumers to perform about the public health costs of the fuel were left in a cloud of smoke as teenagers and housewives alike sped down Woodward Avenue in Detroit and Market Street in San Francisco. If any driver fretted about the mundane concern of low-dose lead poisoning, their worry was incinerated in the fiery engine of American enthusiasm.

There would be no stopping it. By July 1926, barely a year after Hamilton's emotional plea for American vigilance against such an obvious danger, she could tell the tide had irreversibly turned. She felt she was left to battle a runaway bus with nothing but numbers and papers. Science was no match for a society remaking itself around the automobile.

Despite losing the fight to which she had committed so many years, she did eke out two small wins that would prove meaningful over the long term. Since the data in the study had shown notably higher blood lead levels among garage workers and drivers who used Ethyl gasoline versus those who didn't, she convinced Cumming in several letters that the government needed to at least protect those workers. Cumming hemmed and hawed for several weeks about his power to regulate and how any such requirement would be enforced. But he finally relented to several rules: Ethyl could only be manufactured and mixed in small quantities, and warning labels must be put on any pipe or container in any facility that made, processed, or sold tetraethyl lead. Also, based on Hamilton's suggestion, Cumming agreed to require Ethyl gasoline to be permanently colored red to distinguish it for workers or buyers who *were* mindful of avoiding it. In a long career of fighting for the unassuming and marginalized, these were small wins, but victories nonetheless.

After the manufacturing process for tetraethyl lead incorporated Hamilton's basic safety precautions, worker deaths slowed. Ethyl also grew more adept at paying off accident victims quietly, thereby quashing unfavorable press coverage.

The spoils of the fight, of course, looked much different for Ethyl Corporation. Americans bought nearly 10 billion gallons of gasoline in 1926, a bigger number than ever before. Every day, a higher proportion of it was Ethyl. In the decade to follow, annual gasoline consumption in America would rise further to 12 billion gallons, and then to 19 billion (today it's more than 130 billion gallons). Before long, the three-cent royalty on every gallon of Ethyl amounted to more than $300 million.

Ethyl had several more hiccups along the way to its domination. Heavy users noticed it corroded their fuel valves after a few thousand miles. New York City still refused to lift its ban on selling Ethyl gasoline. And in late 1925, a Navy airship fueled by Ethyl earned the company a black eye when it broke apart in a storm above Ohio.

But these were small matters compared to the public health controversy now in the company's rearview mirror. They were also matters that Kettering and Midgley felt they could easily address, either with chemical tweaks for the valve issue or with clever public relations funded by the quickly mounting royalties.

Inside Ethyl Corp., one of the fiercest ongoing debates in the late 1920s was whether to continue exclusive contracts with Esso, American, and Gulf refineries or whether to make Ethyl gasoline available to any refinery that wanted to make it. Kettering thought that keeping known partners would ensure high quality and maintain exclusivity, like selling fine wine. But Irénée du Pont, the president of DuPont, pushed for a "free-selling" policy that would license Ethyl like any industrial chemical and, in effect, put the fuel in the maximum number of cars. After gasoline sales plummeted during the Great Depression, the Ethyl board sided with du Pont, and, by 1933, Ethyl was everywhere, often with no brand name at all. "Ethyl gasoline" was simply becoming "gasoline."

Hugh Cumming had one final act to play in this story. He had already completed his duties as America's top health official in holding the 1925 public health conference and then supervising the independent committee's rushed research of Ethyl gasoline.

But he still felt he had responsibilities to Kettering, Midgley, and the remaining officials at Ethyl since, fair and square, the committee found their product suitable for sale. They weren't formal obligations, nor were the men close friends, but they were all white men in the boys' club of power brokers in the 1920s. And there was one distinct way that Cumming could help a successful American company continue to succeed.

Once Ethyl found explosive market share in the United

States, in 1927 and 1928, the company turned its focus, and its eagerness for profit, to the rest of the world.

Ethyl could manage almost any aspect of exporting a product overseas. The company had relationships with three of the biggest American corporate names—General Motors, DuPont, and Standard—and contract partnerships with a dozen more refiners and distributors, many with footprints in other countries. Under normal circumstances, all a company needed to do to enter a foreign market was to license its production to a local manufacturer and maybe set up an office and buy a few ads to build brand recognition. Senior Ethyl officials could make periodic visits to ensure quality and collect royalty checks on their way home.

This was the general route taken by many expanding American companies in the 1920s, including brands still unknown to the world, like Coca-Cola, Procter & Gamble, and Goodyear. Anyone in the automobile industry could look to the Ford Motor Company as an example. In the teens and twenties, Ford was expanding into the U.K., Germany, and Argentina, selling cars that, conveniently for Ethyl, also needed fuel.

The only problem was that Ethyl's public health controversy followed it abroad. Even if news of the Bayway incident hadn't spooked readers of the *Manchester Guardian* and *Le Petit Parisien*, each country's health boards and university scientists were familiar with the controversy that had, for a time, engulfed the United States. And if they had never heard of Ethyl, they could fall back on the knowledge that lead in any form had a long history of sickness and death.

Ethyl made its first international expansion in 1927 with partnerships in Canada. The process was smooth, in large part because of the proximity of Detroit to Ottawa, which allowed Earle Webb, the president of Ethyl Corporation, to make occasional trips to the Canadian capital to personally convince Canada's Department of Health that Ethyl was perfectly safe, and he had a U.S. government report signed by Cumming to prove it.

A year later, Ethyl found it harder to break into the U.K. A group of scientists in London had assembled into the same loose collective that Hamilton, Henderson, and other U.S. scientists had formed to oppose Ethyl. One of them, a government surgeon named Norman Porritt, penned an article in the prestigious *British Medical Journal* that went a step further than any U.S. scientist ever had. Citing his own bouts with mysterious pains and fatigue after drinking water from a well contaminated with lead, he declared that there was a new type of lead poisoning the world should awaken to. It wasn't the high-dose kind that caused instant insanity or death but a different kind that came on extremely slowly, caused by "infinitesimal doses of lead extending over a long period of time." A group of British scientists used Porritt's research to petition the British Ministry of Health to reject Ethyl gasoline.

American leaded gasoline might have hit a wall in Europe if Ethyl's president Webb had not sent an urgent telegram to Cumming complaining of an impending unfavorable decision in London.

As a personal favor, Cumming wrote a letter to his counterpart, the head of the British Ministry of Health, explaining that Ethyl had been given a full scientific airing in the United States and there were "no good grounds" for prohibiting its use.

Cumming vouched for Ethyl gasoline, and his credibility, as an officer of the government, was considered sufficient. Ethyl started appearing in U.K. petrol stations a year later.

Cumming wasn't corrupt. He wasn't getting kickbacks or bribes. He was simply caught in the wave of a successful American product. Cumming could clearly see that Ethyl was the literal fuel powering the country's economic expansion and that, after the Great Depression, Ethyl would be the fuel to help the country bounce back. If there was any awkwardness, it was in the way that Cumming's agency, the Public Health Service, sat within the U.S. Treasury, which judged all health questions through the lens of the nation's economy.

Ethyl enjoyed this government stamp of approval and used it whenever new roadblocks appeared. Webb asked Cumming to persuade New York City to drop its ban on Ethyl, and when he did in 1928, Ethyl went back on sale in Manhattan. Over the next several years he sent warm letters endorsing Ethyl to the governments of Switzerland, Australia, New Zealand, and South Africa, as well as dozens of U.S. state legislatures still wavering about the fuel. As the list of adoptees grew, the sales pitch became easier. Cumming appeared to be saying, *If so many countries are clamoring for this American-made product, how dangerous could it be?*

Such were the conditions that helped Ethyl conquer the world. There were still discussions in some countries about iron carbonyl and other anti-knock alternatives to tetraethyl lead. But by the mid-1930s, Ethyl had become a runaway train. Its economy of scale made it cheaper to produce than almost any other type of blended gasoline. Ethyl Corporation exploited this by dropping the price of the raw materials lower and lower, reducing the cost of tetraethyl lead from fifty cents per cubic centimeter in 1929 to just eighteen cents in 1942.

Even with the price drops, every party in the gasoline supply chain wanted to keep America's engines purring. Because every part of the gasoline supply chain—from Ethyl Corporation that owned the patent, to the oil companies that made the fuel, the automobile companies that sold gas-guzzling cars, and the governments that collected tax revenue on every gallon sold—was making more money than ever before.

# Part Five

# PHASEDOWN

# 17

*Thomas Midgley's sprawling mansion and award-winning lawn in Worthington, Ohio*

## Worthington, Ohio — 1927

In 1927, Thomas and Carrie Midgley bought a fifty-acre plot of land in Worthington, Ohio. Worthington was just north of Columbus, where Midgley grew up, and he wanted to build a bespoke mansion somewhere flat and expansive.

It helped that he was rich, and getting richer by the day. The financial details of his contract with Ethyl Corporation have never been made public, but—judging from the way the company regularly credited him for inventing its namesake product—it's reasonable to assume that in the first few years of Ethyl's growth he made hundreds of thousands of dollars, and possibly millions.

It also helped that he no longer had one job but three. In 1925, he turned down an offer to be a highly paid vice president of General Motors. Thinking inventively, Midgley crafted an even more lucrative arrangement as a consultant to GM, a senior researcher for Ethyl Corporation, and a board member of the newly formed Ethyl-Dow Chemical Company that was created to produce and market tetraethyl lead.

Without saying it, Midgley's bosses were relieved to see him take a step back from the company's labs. Midgley could no longer be managed—not by his bosses or even his wife—and his erratic behavior had grown almost manic. He had become overconfident and possessed by his own thoughts, like a true mad scientist. He worked late in the evenings and early in the mornings, sometimes calling colleagues in the middle of the night to share a new idea. He lost track of chemical samples so often that his exasperated colleagues lobbied Ethyl to build Midgley his own lab.

Midgley loved to drive in those years. Many days he left Columbus at 5:00 a.m. and arrived at General Motors in Detroit by 9:00 a.m. He worked a full day and returned home by 1:00 a.m. Early the next morning he was back at his desk, writing reports and answering letters. Midgley had no time to feel exhausted. He was possessed by the urgency of his work, as though he felt he was running out of time.

Such an intense focus on himself—and only himself—caused his marriage to slowly unravel. Carrie grew tired of her husband's heavy drinking and constant disappearances. Sometimes after arguments he would leave for days at a time to blow off steam in New York, Detroit, or Chicago. When his son asked where he was going, he told him he was doing his "hat trick," in which he'd put on his hat and walk out the door.

Midgley was famous and he knew he was famous. Not as famous as Charlie Chaplin or Babe Ruth, but a bona fide celebrity of science. He drank his own elixir. No one could tell him what to do.

The only thing preventing Midgley's colleagues and his wife from casting him off entirely was that he was still a gifted chemist and inventor, one of the country's best, and raking in boatloads of money. Midgley was thirty-eight and still had the energy and enthusiasm to solve big puzzles.

He also still had a direct line to Kettering, who after the Ethyl affair was vindicated and rewarded with a new job: head of all research for General Motors. Kettering was the only person Midgley saw as an equal, and the two would shoot the breeze at all hours.

In 1928, one of his roundabout conversations with Kettering launched Midgley on another quest of discovery that would prove consequential to the world.

In addition to producing cars, General Motors also owned a refrigeration company called Frigidaire. Like Ethyl, Frigidaire was born in the labs of Kettering and Midgley's former company Delco in Dayton before it was brought under General Motors. Cool air was something that all cars lacked, including Ford's cars, which made General Motors see a distinct advantage in developing air-conditioning compressors in automobiles.

Air-conditioning was still mostly a theory because no one had figured out how to make the simple idea behind it work in practice. In principle, when a liquid evaporates into a gas, it absorbs heat and lowers the temperature around it. Making consistently cool air meant repeating this process over and over to compress the gas back to a liquid so it could evaporate again and again.

For it to work, however, the system needed a chemical compound that was stable enough that it could be recycled thousands of times inside a closed loop system without losing its potency. The best candidate so far was sulfur dioxide, but sulfur dioxide was a toxic chemical that smelled so foul that the odor was known to wake up a person in a deep sleep and send him racing for fresh air. Leaks of sulfur dioxide and other refrigeration chemicals killed so many people in 1928 that the media labeled refrigerators "death gas ice boxes."

Kettering wanted Midgley to hunt for a better refrigerant chemical than sulfur dioxide the same way he had once hunted for tetraethyl lead.

This was a challenge tailor-made for Midgley, who still carried the periodic table in his breast pocket. He was by now so fluent in the language of chemistry that he regularly had dreams about chemical compounds. He could recall with precision which elements would bond together and list from memory their combined qualities.

His advanced knowledge made this new quest exceedingly easy. And unlike the search for tetraethyl lead, which consumed Midgley for years, he found the new refrigerant in just a few hours.

First, Midgley pulled out his crumpled periodic table and let his eyes dance over the letters. He needed a combination of elements with extremely rare qualities, like trying to thread a needle with nylon rope. It needed to be chemically stable and nontoxic, have a low boiling point, and not corrode the pipes of a refrigeration system. It also needed to be widely available at low cost.

His eyes went to fluorine, which seemed like it could be a strong candidate except for its toxicity. He wondered if it could be made less toxic by bonding it with other elements. His mind raced and his pen fluttered, and before long "everything looked right," as he later put it. He had come upon a new type of theoretical compound combining fluorine, chlorine, and carbon. This class of compounds would be nicknamed "chlorofluorocarbons," later shortened to "CFCs." One of the first CFCs that Midgley tried became Frigidaire's top refrigerant.

Midgley had done it again. The solution to a problem that vexed other chemists came easily to him. With Kettering's pushing, CFCs would follow the exact same path as Ethyl gasoline. Frigidaire partnered with DuPont to produce and distribute a strong CFC called Refrigerant 12 (or R-12 for short), giving it the brand name "Freon." To oversee the entire operation they started a new firm, called Kinetic Chemicals, where Midgley was again made a vice president.

Midgley enjoyed demonstrating how the new compounds worked. He proved how they were apparently nontoxic and nonflammable through an exhibition similar to his 1924 demonstration when he breathed in the fumes of tetraethyl lead to prove it was harmless. Several times in 1930 he wowed audiences by inhaling the vapor of Refrigerant 12, holding it in his mouth and then exhaling it slowly onto a candle that flickered out. The crowd loved it.

Within a year, Frigidaire was selling Freon-filled refrigeration units to cool railcars, movie theaters, restaurants, factories, and schools. A few years later General Motors marketed that "air-conditioning," as it was dubbed because of how it *conditioned* the air to remove humidity, could reduce a car's inside temperature as much as ten degrees.

By 1935, the makers of household refrigerators started using CFCs too. And when they did, the effect was more world-changing than Midgley could have imagined. It was suddenly possible for millions of families to keep food fresher for longer. This meant weekly savings on groceries and fewer trips to the grocery store, but even more dramatic was a reduction in food poisoning and bacterial disease. Cheap refrigeration transformed medicine as well, making it possible to preserve vaccines long enough to reach people outside cities. In the decades to come, vaccines for polio, tetanus, influenza, typhoid, hepatitis, and dozens of other maladies would prevent millions of premature deaths. By 1950, this improvement alone helped extend the average American lifespan by almost ten years.

Coldness was a pleasure, and a much bigger joy than Ethyl gasoline had brought. It changed people's lives in simple but dramatic ways and weakened old-timey industries like milk delivery and icehouses. The biggest pleasure was ice at home, which formerly had to be delivered in giant blocks cut from frozen lakes in the winter and wrapped in flannel in insulated warehouses. Refrigeration allowed soda fountains and bars to serve cold drinks. Corner stores could sell ice cream all year.

This was a turning point in human history. And the gravity of these improvements left no room for objections or even light concern. Unlike Ethyl, there were no health experts concerned about CFCs. There were no doctors who warned about the future damage that CFCs would do to human bodies or to the planet. CFCs did not have the historical reputation of lead or the abundant number of health studies by the nation's top universities. There were no industrial accidents, no sick workers, and no people killed by the newest refrigeration chemicals.

But just as with tetraethyl lead, the true effects of CFCs wouldn't be clear for decades. Because it was impossible to foresee at the time that the main strength of CFCs—that they were extremely stable—would turn out to be their most dangerous attribute.

Almost fifty years passed before British scientists noticed something strange happening in Antarctica. The thick layer of ozone ($O_3$) molecules that protected the earth from ultraviolet radiation seemed to be thinning. It took them longer to figure out that Midgley's stable CFCs were in fact *so* stable that they resisted breaking down for years until they had floated their way to the upper levels of the planet's atmosphere, where the sun's radiation finally broke them apart. Once broken, the chlorine atoms became so reactive that they bonded instantly with ozone molecules and destroyed them. The effect was cascading. A single chlorine atom could break apart thousands of ozone molecules and then regenerate to attack even more of them.

Two scientists alerted the world to this phenomenon in a 1974 edition of the journal *Nature*. At first, DuPont ignored the findings. But when mass media reports began to worry the public, who were already jittery about atmospheric attacks during the Cold War, pressure grew. DuPont responded by shifting the burden of proof the same way it (and Ethyl Corporation) had done with tetraethyl lead. DuPont executives pledged that if scientists could prove that CFCs were a threat to public health, they would stop making them. Until then, the company had not seen data that warranted shutting down one of its multimillion-dollar revenue streams.

Scientists took the challenge and, using public funds and grants from concerned donors, they performed hundreds of studies on CFCs and the earth's ozone layer in the 1970s and early 1980s. DuPont grew increasingly opposed to this research and argued that the criticism was "not based on authoritative evidence." It took out full-page newspaper ads belittling adversarial scientists and arguing that the chemicals did not pose a provable threat. But the company eventually realized it was outmatched, not by the scientists but by the thousands of lawsuits it could expect from millions of new cases of skin cancer and cataracts in the southern hemisphere linked to its product.

In 1987, CFCs became the rare—and the first—environmental issue that all members of the United Nations agreed was dire and urgent. A meeting in Montreal that year produced a unanimous treaty requiring the phaseout of all CFCs responsible for ozone depletion. The phaseout would happen over decades and remains ongoing as the ozone hole gradually recovers. According to a UN assessment, the ozone layer is expected to be back to its normal state by 2066.

Despite the damage caused by CFCs, DuPont escaped the crisis without any major liability or financial loss. When the company was forced to give up Midgley's R-12 refrigerant, it pivoted almost immediately to another refrigerant *also* created by Midgley called R-134a. The R-134a compound was from a different class of chemicals called hydrofluorocarbons that did not contain the damaging element chlorine.

HFCs, as they're known, are still used. They do not erode the ozone layer, but they do create a powerful greenhouse effect in the earth's atmosphere. In 2021, the Environmental Protection Agency started requiring reductions in R-134a in favor of newer refrigerants that contribute less to global climate change. Even today, many of the most common refrigerants are still dozens if not hundreds of times more potent greenhouse gases than carbon dioxide.

DuPont's final act in the saga over CFCs was to launder the reputations of everyone who was involved in creating a life-changing but ultimately dangerous form of pollution for more than half a century. Beginning in 1987, the company spent millions of dollars and used its high-level connections in government and media to frame the mandated end of its dangerous product—which it had fought against for a decade—as an honorable and voluntary act of self-sacrifice to solve a worldwide problem. Citing its "early and active support" for scientific research on the ozone layer, the company announced it would begin a "total phaseout" of CFC production. Its executives sought to wash away years of opposition by congratulating themselves for doing the right thing. "I know I'm doing something that's important and it felt good," the director of DuPont's Freon division, named Joseph Glas, apparently told his children, and later repeated to reporters.

The effort worked. On April 10, 1988, the *New York Times* reported without any irony that DuPont's corporate responsibility broke new ground and demonstrated that, when asked to choose between their monetary interests and the health of the broader world, corporations could be counted on to do the right thing. "It shows that corporate America can take steps to protect the global environment," the paper stated.

In Washington, D.C., a senator from Montana applauded DuPont and its leaders for their initiative. He called the company "very responsible."

Midgley finally got around to building his mansion in Worthington in 1928. The intensity of his work and constant travel had worn him down. His heavy drinking didn't help, nor did his strange pains and shortness of breath. He told people the house would be a retirement home for his aging father. But Midgley was forty and in poor health, and his friends could tell it was for him too.

Midgley envisioned a 10,000-square-foot Colonial mansion with eight bedrooms, each with its own bathroom and walk-in closet. That alone would make the house remarkable. He also wanted central air-conditioning, an outdoor swimming pool, and a six-car garage for his growing vehicle collection. It would be a temple to Midgley's inventions.

The house at 382 West Wilson Bridge Road was mostly done by the fall of 1929 when the stock market crashed and sparked a nationwide depression. Midgley and Carrie were content in their new house, although not with each other; they were now sleeping in separate bedrooms.

The depression was a deflating time, even for someone as rich as Midgley. Despite the satisfaction of a successful career, a custom mansion, and more money than he knew what to do with, Midgley couldn't help feeling sorry about the pictures in the newspapers of breadlines and destitute workers. He was self-centered but not always selfish. As the nationwide depression grew worse, Midgley decided to fund a public works project of his own. He hired fifty men to constantly improve his estate. They were paid to build roads, dig trenches, and plant and tend to endless flowers and shrubs to beautify the property.

Once the estate reached a terminus of improvement, Midgley had an even more eccentric idea. He had been obsessed with anthills since childhood and wanted to make his own underground world in a bluff behind his mansion. He hired dozens of men to dig two large tunnels that would branch off into underground rooms and chambers before eventually meeting in a giant room with a vaulted ceiling. There was no shortage of men willing to sign up for the strange project, and no deficit of Midgley's money to pay them.

Working all day, and sometimes at night, too, the men dug five hundred feet of tunnels. They paved the walkways with limestone slabs from a nearby quarry and decorated the passageways with wrought iron gates, hinges, and candle sconces, making the lair look like it was from medieval times. People accused Midgley of pouring good money—at a time when most people had little—into a literal hole in the ground. But Midgley didn't care. It was his way of sharing his wealth. Plus he loved the way the tunnels impressed his guests, who marveled at an estate unlike any they'd ever seen.

When the tunnels were done, Midgley turned his obsession to more eccentric hobbies. He recorded himself reading his favorite poems and wrote short stories. He was especially devoted to his lawn. He paid gardeners to plant eight acres of grass around his house and had it cut almost daily at putting green length. When earthworms turned up tiny piles of dirt, Midgley installed sprinklers to drown the worms. And when warm breezes blew in the night and dried the grass, he wired a windmill to trigger an alarm clock to wake him up to dial a telephone code that would restart the sprinklers. The grass was so green and smooth that golf club chairmen traveled to see it. Eventually, a representative for the Scott Seed Company informed Midgley that the company wanted to put a photograph of Midgley's house and lawn on seed packets and on the company's letterhead too.

Midgley carried on like this for most of the 1930s. He grew so manic about his various projects, often while fueled by alcohol, that it became difficult for him to carry on his prior relationships. His children were off at boarding school and Carrie could hardly stand him. Even Kettering took longer to return his calls.

In the fall of 1940, Midgley was in Detroit at a meeting of the American Chemical Society and began to feel tingling in his legs. He drove back to Worthington and saw a doctor, who diagnosed him with poliomyelitis. Polio, as it was known, was a virus that attacked the spinal cord and often led to paralysis.

Midgley didn't believe it at first. When another doctor confirmed it was polio, Midgley spent hours in bed calculating the odds of a fifty-one-year-old man catching what many people considered to be a children's disease. "This comes out to be substantially equal to the chances of drawing a certain individual card from a stack of playing cards as high as the Empire State Building," he wrote to colleagues. "It was my tough luck to draw it." To feel even sorrier for himself, he calculated that dying of a coronary thrombosis was 15,000 times more likely than dying of polio, terminal cancer was 12,500 times more likely, and a fatal car wreck 7,562. He cited no evidence for his calculations. Owing to his scientific renown, his rough calculations about his unfortunate condition were published in the *Journal of Industrial & Engineering Chemistry*.

But Midgley was wrong. Polio was not a children's disease, as he suggested. It was destructive in young people, but it also affected people of all ages—even the fifty-eight-year-old president of the United States. Nor was he on the bum end of long odds. Almost 10,000 Americans caught polio in 1940, 650 of whom resided in Ohio.

Only decades later would historians notice something strange about Midgley's diagnosis. Midgley had polio's common symptoms, including muscle weakness in his legs, as well as persistent fevers and gastrointestinal pain. But of the dozens of doctors Midgley saw in those years, it's curious that none of them—or Midgley himself—made the connection that his neurodegenerative effects were also the same symptoms to be expected after decades of working with industrial chemicals, and especially lead.

By 1941, Midgley was paralyzed from the waist down. His days of driving were over, as was his marriage. He and Carrie were still legally wed, but she was so worn down from two decades of her husband's never-ending intensity that she declined to be his nurse.

As Midgley faded, he devoted his inventive mind to one final creation. To enable him to move his half-limp body, an assistant helped him build a series of crossbeams and ropes to support a harness that would transfer him from his bed to a wheelchair.

He lived this way for several years, growing increasingly frustrated but determined to keep his mind active. In 1941, after the attack on Pearl Harbor, he spoke to U.S. military leaders, using an elaborate setup of microphones in his bedroom, about adding more tetraethyl lead to American and British warplanes. In 1942, he addressed the National Inventors Council via the same system. And in October 1944, while lying horizontal, he delivered his final speech to the American Chemical Society, which elected him president that year as his swan song. He used the address to discuss his final grand theory: that all of history's great men peaked before they turned forty. Mozart, Galileo, Marconi, Edison, Beethoven, Bell, and Bessemer—they, just like him, made their great progress in their twenties and thirties. It seemed like he was trying to rationalize, to himself and the world, that at fifty-five his work was finished.

Thomas Midgley killed himself on the night of November 1, 1944. He concocted a plan to end his torturous existence using the only tools he still had. When Carrie walked into her husband's bedroom the next morning, she found him strangled by the ropes of his contraption. The medical examiner inspected the scene and noted on his death certificate "164a," the medical code for suicide by asphyxiation.

Carrie sold 382 West Wilson Bridge Road a few months after Midgley died. If she grieved him, she grieved quickly. She wanted a fresh start and moved to California.

Midgley's colleagues held an elaborate funeral for their friend. Kettering gave a eulogy and said he considered Midgley "like a son or a brother" and that he left "a lot behind for the good of the world." Kettering and the others agreed that, as a final tribute to their fallen friend, they would spare Midgley the shame of dying by his own hand. In the months and years to follow, all public mentions of Midgley by General Motors, Ethyl Corporation, and the American Chemical Society referred to his death as an accident.

Midgley's legacy would be decided long in the future, but

his prized mansion had less time. A young couple bought the house in 1945. They thought the large rooms and sprawling greens would be a favorable place to raise their five sons. Midgley's flourishes slowly faded. His green lawn turned brown. His tunnels became a hangout for local teenagers.

Finally, in 1964, the state of Ohio saw a better use for Midgley's old estate. Columbus needed to accommodate the thousands of cars traveling daily between the growing suburbs of Dublin and Westerville. There were more of them than ever before, and almost all were running on tetraethyl leaded gasoline. The city wanted to build a circular highway, called the outerbelt, and the old Midgley house was in the way.

With little fanfare, workers arrived one morning, filled Midgley's tunnels with cement, and bulldozed his house to the ground.

# 18

*Alice Hamilton at her home in Hadlyme, Connecticut,*
*in 1957, at age eighty-eight*

## Hadlyme, Connecticut — 1932

Hamilton decided to take time off in the spring of 1932. It had been a hectic few years and she was ready for a break beside the Connecticut River, watching the ferryboats. She also was working on a new book, her second, and needed time to focus.

For the first time in her life, she would spend a season alone.

She was now sixty-three and had only ever lived in communal quarters, first in her family's mansion in Fort Wayne and later in the dormitories of Hull House, the Boston brownstone with the Codmans, and in hundreds of hotels and guesthouses along her endless travels. None of her sisters thought she could withstand the solitude and silence, but she resolved to prove them wrong by keeping a strict routine.

She awoke each morning at seven and went to pour coffee, then ate the same breakfast of fried bacon and toast. She'd go outside to see what the day was like, taking time to notice the bloodroot blooming and the ferns and lilies throwing up sprouts. She would work from nine o'clock until one and then answer letters until three. She spent the afternoon in the garden spreading manure or planting seeds, followed by supper at seven and bed by nine thirty. Her sisters underestimated the sheer force of her self-discipline.

After the Ethyl affair ended, Hamilton buried her disappointment in new work. Just as she had done in her thirties and forties, she spent her fifties taking on almost every project in which her expertise was wanted—and found that the list of requests never grew shorter. She joined the Birth Control League of Massachusetts to advocate for contraceptives. She was called to Washington to testify on the medical merits of abortion. In 1930, President Herbert Hoover invited her to join a new government program known as the President's Research Committee on Social Trends that would document changing behaviors across the country. Like so many times before in her life, the committee had five male members and, for appearances' sake, needed a woman.

The most surprising request for Hamilton's attention came from the League of Nations, the international forum based in Switzerland. Most Americans hated the League, or at least distrusted it on the grounds that foreign entanglements seemed likely to cause more war, not less. But the League asked Hamilton to be on its Health Committee—a less political wing—in hopes of combining the best global minds to fight devastating diseases like cholera, plague, typhus, and smallpox. Hamilton was again the sole woman invited.

On her first visit to Geneva, she learned that white doctors had to act gingerly in giving advice to poor countries about how to fight malaria or influenza—or as she put it, "We have to approach governments tactfully and suggest that perhaps some expert advice might be helpful." Colonial abuse had left its mark of resentment in western Africa and Central America, but by 1928, things had gotten so bad that nearly every country begged for Hamilton and the League to help. Infant mortality rates were sky-high across Africa, malaria had blanketed most of Asia, and Brazil was struggling with leprosy. The native tribes of the Pacific Islands pleaded for help to stop imported diseases like measles, influenza, and dysentery. And Greece was in full panic that dengue had spread to almost every person in Athens. In every case and every country, Hamilton observed that the combined force of hospitals, doctors, and government officials appeared powerless to protect people.

It was an important revelation. For years she believed that her own government was hopelessly flatfooted on basic measures of public health. Yet here was proof that the United States was rather ahead of the rest of the world by a wide margin.

She made two more trips to Geneva. After each one she required days of rest from the intensity of tragedy that confronted her. When her term ended in 1930, she had developed a conflicted perspective on the world. The world was slowly moving forward, and yet there was an incredible amount still to do.

Hamilton finished her book in 1933. It was a tome—nearly four hundred pages and not an ounce of fluff. From the first word to the last she gave a patient and thorough discussion of every elemental toxin known to humanity, including mercury, magnesium, beryllium, nickel, and, of course, lead. She devoted five pages to tetraethyl lead specifically and showed a still-simmering frustration at its reckless use and widespread adoption. "When it was first produced and blended with motor fuel, there was general ignorance of its dangerous nature and of the ease with which it is absorbed by the human body," she wrote.

The book, called *Industrial Toxicology*, came out the next year. It sold poorly, likely due to its clunky name and the relative youth of the field it covered. But fourteen years later, in 1948, a younger female toxicologist named Harriet Louise Hardy, who was inspired by Hamilton to go into toxicology and become a professor at Harvard Medical School, had the idea to revise Hamilton's book and promote it as a textbook. *Hamilton and Hardy's Industrial Toxicology* was published in 1949. It became the definitive guide to known poisons encountered in factories, warehouses, and heavy work sites around the world. As of 2015, it had been revised six times and expanded to over thirteen hundred pages.

Around the same time that Hamilton's second book appeared, however, her first book was starting to cause her trouble. In the summer of 1932, the heads of five companies that sold fire extinguishers wrote to the publisher, Macmillan, to complain about Hamilton's 1925 book, *Industrial Poisons in the United States*. In the book, she labeled carbon tetrachloride, the main agent in fire extinguishers, as a toxic respiratory irritant. The industry lost a major contract with the New York City subway for thousands of fire extinguishers after the subway cited Hamilton's book as its reason. The industry bosses wanted Hamilton to withdraw the claim, and if she wouldn't, then they wanted Macmillan to pull the book from sale, and if *they* wouldn't, then the companies would bring a lawsuit.

Hamilton spent her entire career battling companies that demanded that she cast a gentler light on their businesses. She had seen it hundreds of times in dozens of industries.

But after a lifetime spent politely compromising to avoid conflict so that she could maintain her standing and continue her work, she had finally had enough.

Hamilton hired a lawyer to fight back. The lawyer suggested she simply amend the book to satisfy the executives and be done with it. Who cared about a book published seven years ago? he asked.

She said no.

It bothered her that no matter how hard she worked and how senior in her field she had risen, she, and not male researchers, was always asked to compromise her work for someone else's commercial gain. Yandell Henderson and Howard Haggard at Yale had also published about the toxic danger of carbon tetrachloride, and the U.S. Bureau of Mines had conducted a study arriving at the same results. And yet she, and not the men, was being threatened with a lawsuit to blunt her warnings.

As she saw it, the fire extinguisher executives were making an example of her because she was the "weakest of adversaries," an easy target. And if she was forced to back down, the industry could cast doubt on the other scientists' findings.

She responded to her lawyer that, no, she would not let it go. "I find it hard to continue to be the only victim of this attack . . . all this effort to alter the text of my seven-year-old book and to have me write an article favorable to them is so futile and foolish," she wrote. "It is not simply a question of frightening one woman into submission . . . even with me demolished the facts would remain as they are."

She was fed up. But she was also aware of her limited power against a well-funded industry run by uncompromising men. As the weeks went by and the threat of a ruinous lawsuit grew closer, she had no choice but to agree to an undignified compromise: She would not remove her criticism about carbon tetrachloride, nor would Macmillan pull the book. But she would write an article offering more vague critiques of the fire extinguisher substance that could be interpreted more favorably and help the industry bid for another contract with the subway authority.

She hated this. But she was smart enough to know her options were limited.

"I might lose my temper," she told her lawyer. To another friend she wrote, "I am getting old and garrulous, saying things I ought to have kept quiet."

When she sent her article to be published in the *Journal of Industrial & Engineering Chemistry*, the journal of the American Chemical Society, she was ready to forget the whole affair. But when it appeared in print in May 1933, her anger boiled over again. Without her knowing, the editors arranged for a counterpoint article to appear beneath hers. It was written by a representative for the Chemical Fire Extinguisher Association and referred to Hamilton's work as "misleading and inaccurate" and called into question her credentials and credibility. The dueling articles gave the appearance of debate, the same strategy employed by Ethyl to confuse the public.

The subway authority accepted a new contract and installed thousands of fire extinguishers in stations around New York. Only in the 1950s, after twenty years of use, were carbon tetrachloride fire extinguishers confirmed to cause respiratory disease, kidney disease, liver damage, and various types of cancer. They were phased out in the 1960s and banned by the EPA in 1970.

For Hamilton, the entire incident held one final insult. By sheer coincidence, in the same issue of *Industrial & Engineering Chemistry* where her article appeared was another article by the famous industrialist Charles F. Kettering. Kettering could not have known about the fire extinguisher affair or that Hamilton's article would appear in the same issue, but his article, which extolled the virtues of chemistry, had a taunting bluster all the same.

In the span of three pages, Kettering praised "the possibilities of progress" and wrote that nothing could hold clever industries back as long as they barreled forward over all obstacles. "I have never lost my confidence in . . . American industry," Kettering gloated. "We have only to believe that we are right, we have only to feel that we can do things, and that very feeling will automatically adjust us to the ability to do it."

Hamilton took two of the most memorable trips of her life in her sixties. She was invited to Germany by a group of doctors as part of an exchange program for doctors to observe conditions in each other's countries. She knew that Berlin in 1933 was best avoided. A new chancellor was coming to power and she had read in American papers about kidnappings, beatings in the street, and violence against Jews. But she believed that sharing medical lessons could create common ground between the countries and possibly offer the German people an off-ramp from authoritarianism.

Almost as soon as she arrived in Germany, she noticed the slow breakdown of a proud society: Teachers were censored in their teaching, the government mandated boycotts of certain businesses, and families were being evicted from their homes. She sat for hours on the base of the Brandenburg Gate and watched an endless procession of children marching and singing devoutly of their new leader. She heard Adolf Hitler give a speech, and then the next day she watched as he dissolved labor unions and nationalized entire industries.

It seemed bizarre to her how many German people

supported what was an obviously cruel regime. One man she met recognized her American accent and told her that Franklin Roosevelt was not as strong as Hitler and was probably a Jew. Another woman, a young architect, declared she was ready to drop her career and revert to domestic duties if the "Fatherland" asked her to. Young people seemed especially open to the Nazi philosophy of purity. "It is staggering to me to hear such talk in a country whose intellectual achievements I have held in such high esteem," she later wrote.

Watching Germany fall to darkness was like watching an energetic workman collapse with sickness. That was how poison worked. If not stopped quickly, it could infect broadly and eventually pollute entire countries.

Hamilton was invited back to Germany five years later, in 1938, for a meeting of the International Congress of Occupational Accidents and Diseases in Frankfurt. By then the situation had gotten considerably worse. No Jews were allowed at the conference—which was unfortunate because so many industrial doctors in Europe were Jewish—and the air was filled with constant worry. Hamilton was, once more, the sole woman invited.

Her second visit coincided with the run-up to possible war. Germany's plan to annex Czechoslovakia was on every radio broadcast and shrouded every conversation. When Hitler spoke, the country listened and seemed entirely resigned to what lay ahead. "It was to me a dreadful thing to see those Germans, educated, civilized human beings, listening with no sign of revulsion to the most hateful outpouring of abuse it has ever been my misfortune to hear," she later wrote in her memoirs. As an outsider, Hamilton could see Hitler for who he was: a man devoid of generosity, magnanimity, chivalry, integrity, and pity. A man who spewed lies that he knew were lies. But the German people were entranced in a tribal groupthink and seemed to adore him. Hamilton spent several nights alone, wondering if it had been a mistake to visit Germany at such a dark time and whether she'd ever make it home.

Then, in her final days in Frankfurt, news broke that Hitler had met in Munich with the leaders of Italy, Britain, and France. They agreed to a treaty that awarded Hitler a part of Czechoslovakia home to ethnic Germans. Hitler declared it would be his last territorial claim in Europe. Everyone, including Hamilton, felt the relief of war averted.

But the Munich accords did not last. Five months after Hamilton's visit, Hitler invaded Prague, and six months later German troops entered Poland, which initiated a full-scale war that would encircle the world.

Before she returned home from Europe, Hamilton spent several days in Holland. Hamilton's sisters were eager for her to come home, but the head of a factory in a small Dutch village that made one of the decade's most fashionable products, rayon, invited her to visit. Rayon was a synthetic fiber that resembled silk and was made from recycled plant cells. It also produced a toxic byproduct, carbon disulfide, that caused factory workers to go insane. It was the same storyline but with different characters. Hamilton walked the rows of the factory, as she had done dozens of times before, and jotted down suggestions.

Hamilton retired from Harvard in 1935. The medical school informed her she would become a professor emeritus, which was a gentle enough push for her to understand that after sixteen years at Harvard her time was up. Sixty-six was past the normal retirement age for most workers, but Hamilton felt she had another decade of work in her, possibly two.

Her sisters prevailed on her to write another book, her third; it would be her autobiography. She wrote it on a small

Royal typewriter in her downstairs study at Hadlyme, recalling her upbringing in Indiana, her time in Hull House, her appointment at Harvard, and everything since. When it appeared in 1943, Hamilton believed that *Exploring the Dangerous Trades* would be one of her final pieces of work.

Ten years later, she had amassed another decade of stories. She helped the fur industry protect workers from mercury in felt fibers and walked miles of munitions factories and shipyards during the war to limit sickness among soldiers. When the war ended, she dove into the causes of beryllium disease and the way a new insecticide known as dichloro-diphenyl-trichloroethane, or DDT, seemed to be causing an increased number of cancer cases. In each case, she wrote articles, published papers, and traveled to give speeches.

The FBI started following Hamilton in 1953. She was eighty-four but still vibrant and influential. Her lifelong sympathies for marginalized people drew her to the spreading national communist movement. She was also affiliated with the American-Soviet Science Society, the National Federation for Constitutional Liberties, and the Friends of the Soviet Union, all small groups that were possibly fronts for communists. Perhaps most suspicious, she had a well-known history of pointing out the dark sides of capitalism.

FBI agents trailed her for several months. They read her mail at Hadlyme and established informants in the nearby towns of East Haddam and Lyme to report her comings and goings. They watched as she wrote a letter to President Eisenhower defending the right of the Communist Party to exist. And she wrote two newspaper articles urging the release of twelve communist leaders on trial for activities deemed un-American.

In the end, the Bureau concluded that there was little to be gained from arresting or prosecuting a five-foot-four, 105-pound retired octogenarian—curiously accurate details noted in her classified file. After Hamilton turned eighty-five, the FBI closed the investigation.

Hamilton eventually started to slow down, not because she lacked enthusiasm but because the field of industrial medicine had grown enough that she no longer had to accept every assignment. Conferences began to fill with younger researchers who devoted their careers to specific toxins like lead, asbestos, formaldehyde, and benzene. Those scientists, with greater resources, more advanced tools, and government funding, dug deeper with more complex studies than Hamilton ever had. Hamilton was an elder scholar, a mentor and role model. She advised younger scientists to be "on constant watch for old poisons reappearing in new industries." In the widening field of safety, sickness, and worker protections, she had ascended to matriarch.

At age eighty-seven, Hamilton was named Woman of the Year in Medicine by the American Medical Women's Association. When she heard of the honor, she laughed it off. It came rather late in her career, she thought. But at least for once she could be sure that such an honor—awarded to a woman by other women—hadn't come just because they couldn't find any qualified men.

The award and other recognition had a trickle-down effect on other women. Seeing a woman honored and respected empowered others to go into the sciences and, once there, to speak out on matters they deemed urgent.

One of those younger women, a marine biologist in Maryland named Rachel Carson, followed in Hamilton's footsteps in 1960 when she noticed a curiously high number of birds dying in the northeastern United States. Just as Hamilton had done for more than sixty years in almost every heavy industry, Carson studied the subject, collected data, and eventually published a series of articles in the New Yorker about the damage done by synthetic pesticides. DuPont and other chemical companies that made pesticides spent more than $200,000 trying to smear Carson. They called her a "hysterical woman" and a communist, and they threatened lawsuits.

But something about the country and its regard for female scientists was changing. In 1962, Carson's articles were adapted into a book titled *Silent Spring* that awakened the public to the damage humans were causing to the world. It sparked a movement that led, a decade later, to the phasing out of one of the worst synthetic pesticides, DDT. *Silent Spring* became and remains one of the top-selling nonfiction books of all time.

Hamilton didn't start to think about death until she turned ninety. And when she did, she hoped she would go quickly. The indignities of old age were bearing down. She experienced small strokes so frequently that she learned to live with them, and did the same in 1961 when a gastric ulcer burst and almost killed her.

She joked to friends that she did not recommend longevity. The worst part was losing her sisters and friends. Norah the artist died first in 1945, her cousin Jessie in 1960, and a year later her beloved sister Edith died, leaving behind a long career as a world-renowned classicist honored by the Greek government. Even her brother Quint, who was seventeen years younger than her, met his end in 1967, when Hamilton started what would be a yearslong bedrest. When Quint died, Hamilton and her sister Margaret were the only ones left. None of the Hamilton sisters ever married, which made each death a heavier load for those who remained. Margaret spent years sitting vigil by Alice's bed, not wanting her sister to die alone. But in 1969, at ninety-seven, Margaret died first.

Hamilton once had a dream that she would live to a hundred. In the end she made it one year further. She withstood the strokes for years, but they worked like a poison, eroding the pathways in her brain until it could no longer function. On September 22, 1970, she spent her final moments looking out the window at the fall leaves and the ferry crossing the Connecticut River.

Days later, word of her passing would appear in more than a hundred newspapers around the country. Accolades rolled in that would have surprised a woman who had long shunned public recognition. The *Hartford Courant* called her "an amazing lady" and the *Sidney Daily News* of Sidney, Ohio, reported the death of a "famous doctor." As far away as California, the *Ventura County Star* and the *Chico Enterprise-Record* labeled her a "pioneer in medicine." Her hometown paper, the *New London Day*, marveled how one woman could span so many fields: physician, teacher, social worker, chemist, and feminist. Her legacy, the paper wrote, was "a sterling example of what one determined person can accomplish." When President Nixon heard of her death, he sent a note of condolence to the house at Hadlyme, not knowing who would be there to receive it. Nixon was the last of twenty U.S. presidents during Hamilton's lifetime; more than half of them she had come to know personally. He paid tribute to Hamilton and her "lasting contributions to the well-being of our people and of men and women everywhere."

Nixon's bigger tribute to Hamilton would come several weeks later. In December 1970, Congress did something unusual: It agreed on a set of reforms to protect workers. The law was named the Occupational Safety and Health Act, and when it passed both chambers, it required for the first time that all American employers provide working conditions free from known hazards and mandated that workers could report unsafe conditions without fear that they would lose their jobs. Those who got sick or hurt at work would qualify for compensation. And anyone who worked with heavy machinery or industrial chemicals would have access to training sessions to learn about the materials they were working with and how to manage them safely. To this day, the agency that oversees these reforms considers Hamilton its "founding mother."

In a scene Hamilton could not have imagined in 1897 when she moved into Hull House to work with the poor, Nixon signed the act into law four days after Christmas in 1970. It was the first of six landmark laws Nixon would enact to protect public health, natural resources, and even wild animals.

All her life, Hamilton referred to her work as industrial hygiene and public safety. But Nixon's laws brought her fight into a larger movement—environmentalism—that would endure as the centerpiece of social justice battles long after her lifetime.

# 19

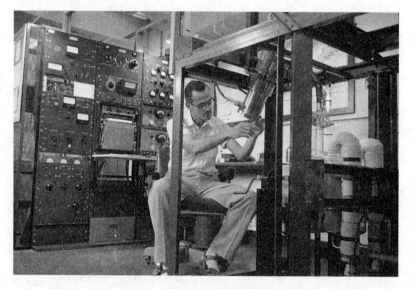

*Clair Patterson in his lab at California Institute of Technology*
*(Caltech) in 1957*

## Los Angeles, California — 1963

O ne morning in 1946, a young graduate student named
Clair Patterson walked into his professor's office at the
University of Chicago. Patterson had come to Chicago to
work under Harrison Brown, a nuclear chemist who had recently helped develop the nuclear bomb.

Patterson needed a dissertation topic for his PhD. He was

young and energetic and up for the usual graduate student drudgery of performing tedious calculations. So Brown suggested that Patterson should figure out the age of the earth.

Patterson had a good laugh at this. Dating the earth was almost preposterously complicated. Until the 1930s, the best that anyone who wanted to date the planet could think of was to add up the generations of characters in the Bible or, to be more scientific about it, perform a heat loss simulation on a ball of magma cooling to the modern temperature of the planet's crust. Not long before, a different chemist at the University of Chicago had discovered a way to date old items like ancient scrolls and human skeletons by measuring the half-life of a radioactive isotope known as carbon-14 ($^{14}C$). But carbon-14 had a half-life of only 5,000 years, which meant there wasn't any datable carbon left after 10 half-lives, or older than 50,000 years.

But there was another route Patterson knew about, at least in theory. Chemists had discovered a strange property of uranium that made it decay very slowly into lead. No atoms were lost in the process, and the decay happened at a constant rate, which meant that if Patterson could find and isolate a piece of lead ore, he could measure how much uranium from the earth's birth was still in it. Accounting for the rate of decay, he could then work backward and arrive at a starting point.

Patterson started to collect pieces of lead ore to feed into a lab machine called a mass spectrometer that would count the lead and uranium molecules in each sample. But when he ran his first sample, his readings were too noisy. Trying to measure tiny amounts of lead was like trying to see a tree in a dust storm. It was there, but there was too much junk in the way.

He tried smaller samples to get a narrower reading, but those were noisy too. He blew on the samples, washed them in distilled water, and built small enclosures around them. Still noisy. He washed them in double-distilled water, then triple-distilled. He covered work surfaces with plastic wrap. He mopped and vacuumed the lab and scrubbed the tables and chairs. Still noisy. He washed every piece of lab equipment and rinsed it in the aggressive cleaning agent potassium hydroxide. All this brought the background levels down. But there was still too much noise.

This was getting ridiculous, he thought, so he removed the samples from the spectrometer to test the background levels of lead in the lab. The reading surprised him. There was not a little extra lead in the room—there was a *lot* more, about two hundred times more than he expected. "I tracked back and I found out there was lead coming from here, there was lead coming from there; there was lead in everything I was using," he later recalled. "It was contamination of every conceivable source."

Patterson began to understand why no other researcher had beaten him to the age of the earth. But what looked like an insurmountable obstacle seemed to Patterson like a challenge. He decided he needed to make his lab the cleanest two hundred square feet on earth.

Over the next few years, he made his cleaning protocol his primary job. He set rules on who could come into the lab and what they could wear. Windows were never to open; vents were taped over. In 1951, Patterson was offered a postdoc job in the geology department at Caltech in Los Angeles. He accepted on the condition that he could design his own hyper-clean lab free of lead. He ordered the removal of lead gaskets used in gas lines and lead oxide putty in window seals. He demanded that the building's lead pipes be replaced with copper, that electrical wires be rerun without lead solder, and that his worktable be replaced with a single piece of stainless steel without screws or gaps that could collect debris. Everyone who entered had to remove their shoes, and eventually they had to strip down to their underwear and don plastic suits.

Finally, after four years of obsessive scrubbing and bossing people around, Patterson had his long-awaited breakthrough in 1953.

He prepared the purest sample he could find: lead from a meteorite found in Arizona that was formed around the same time as the earth. He loaded a piece of the meteorite into the mass spectrometer. And when it reported its reading, he punched the numbers into his uranium-decay equation and . . . there it was: the age of the earth, 4.5 billion years.

Patterson was so excited, he almost collapsed. For the next few days he felt as if he were having a heart attack.

He announced his findings at a conference in Wisconsin several months later. He might have ended his career there, at thirty-one, with one of the biggest findings in modern geology, or he might have spent the next few decades trying to get the planet's age ever more precise.

But something else bothered him. His effort to rid a single room of lead contamination had brought up a troubling question. There was substantially more lead in the world than any scientist would expect. Where the hell was it coming from?

To calculate a rough estimate of how much lead existed on earth, Patterson reasoned that the planet's oceans had been around longer than any land feature, so they should have the highest accumulation of the element. He chartered a boat in Boston and dropped bottles on ropes to collect water samples at different depths. He assumed that the lower depths would have the most lead because of the way sediment settles. But when he pulled the bottles up and brought the samples back to the lab, it appeared that the opposite was true: The upper layers of the water column had more lead than the bottom layers, as much as one hundred times more.

This was a head-scratching finding. It was an expensive

one too. Chartering boats and paying lab technicians was costing more money than Caltech would give him. A friend suggested that the oil industry might be interested in the geochemistry of the deep ocean to help companies locate new oil deposits. Patterson agreed, and after a little letter writing in 1955, he got the American Petroleum Institute, the lobby for the oil industry, to commit $30,000 per year to continue his work.

With the extra money, he came up with an even bolder experiment. If lead was floating around in the ocean, then it was also frozen in ice. Glacial ice is separated by thin layers of summertime dust. If he could count the layers between the dust, then he could measure the lead accumulation in the ice year by year.

Patterson traveled to Greenland to take ice cores. And when he tested the layers, a man who had filled his career with one surprising discovery after the next had the most shocking realization yet. According to the glacial ice, the lead buildup in the air started at a precise moment, in 1923.

It wasn't hard for anyone to put together what happened in 1923. But Patterson spelled it out in a brief summary of his findings that he published in the journal *Nature* in July 1963. The bulk of the lead measured, he and a colleague wrote, "could readily be accounted for as originating from leaded gasolines."

At first, the Lead Industries Association, the powerful lobby group behind the companies that made plumbing, paint, batteries, fishing equipment, roofing materials, stained glass, and circuit boards, thought to deal with Patterson gently. The LIA sent him a contrary study from Stanford that was funded by the lead industry. *See,* they said, *how can you be sure when there's debate?*

When Patterson didn't respond, two petroleum executives paid a friendly visit to his lab at Caltech. Did he enjoy the grant money from the oil companies? they wondered. Because there could be more—*lots* more—if he changed the direction of his research and, say, focused back on the age of the earth.

Patterson wouldn't hear of it. "They tried to buy me off," he later told a friend. Instead of agreeing to an arrangement that would have provided him and his research assistants with enough money for the rest of their careers, he treated the executives to a long lecture about how he believed, but wasn't yet sure, that leaded gasoline had likely covered every inch of land and water on earth with lead dust and contaminated every living organism. A few weeks later, the American Petroleum Institute revoked its grant to his lab. Not long after, Patterson learned that the lead industry was working to block all his other funding too.

When word got out that Patterson could not be managed, cajoled, or bribed, the Lead Industries Association and one of its prominent members, Ethyl Corporation, decided to do the next best thing: It offered generous grants to other scientists to undercut and discredit Patterson's work.

Patterson was not drawn to controversy. A man with weaker resolve might have changed his mind. The reduced funding and the aggressive competition made a meaningful dent in his ambitions for future research and the livelihoods of the graduate students who worked for him. But he felt he had come across something substantial, and when he thought deeply about it—about the implication that every person alive was being unknowingly poisoned—it was too much for him to ignore. Studying, publishing, and warning about lead from man-made sources would be his new crusade.

Patterson spent most of 1965 verifying his calculations. It occurred to him that if he was alleging wide-scale lead poisoning, he should know the precise threshold at which lead poisoning occurs. How much was normal in a person's body, and how much was too much?

This was no longer a matter of geology or chemistry but of physiology, and thus outside his expertise. The best he could do was rely on existing research from physiology experts. As it happened, the nation's top scientist on all matters of lead and human physiology was a man who had now researched lead for more than forty years, named Robert Kehoe.

Forty years earlier, Kehoe was the same scientist recruited by Charles Kettering to boost Ethyl Corporation in 1924. In the decades since, Kehoe served more than thirty years as Ethyl's medical director and also as the director of the Kettering Laboratory at the University of Cincinnati. He performed dozens of studies on the physical impact of lead in the human body and published authoritative findings, all supported by generous funding from Ethyl and the broader lead industry. During that time, Kehoe maintained that he had complete autonomy and followed only the data, but historians have since observed that his research was selective in its hypotheses, experiments, and conclusions. He deduced that lead was "harmless" in the human body under a limit of 80 micrograms per 100 grams of blood. He also proposed that the body ejected harmful lead at a sufficient pace to avoid poisoning. The 80-microgram threshold became known as the Kehoe limit and had gone unchallenged for more than three decades.

Patterson was determined to understand how Kehoe arrived at the 80-microgram number. After working almost around the clock for a year, he realized a central Kehoe assumption was wrong. Lead did not get ejected from the body nearly as quickly as it entered it. And Patterson's ocean research proved that lead was accumulating in the air faster than ever before.

In February 1965, Patterson was invited to submit a summary of his counterevidence challenging Kehoe to the prestigious *Archives of Environmental Health,* a publication of the American Medical Association that would give his work much greater visibility. Prior to publication, as was custom, the journal invited experts in the field of lead physiology to review Patterson's calculations. It would have been strange not to invite the leading expert of them all, Robert Kehoe, to weigh in. When Kehoe reviewed Patterson's work, Kehoe, who was being directly challenged by Patterson, did not hold back. He called Patterson's conclusions "exceedingly bad" and "naïve."

"It's an example of how wrong one can be," Kehoe wrote. "He is so woefully ignorant and so lacking in any concept of the depth of his ignorance that he is not even cautious in drawing sweeping conclusions." In the end, Kehoe said he endorsed the article for publication only so that Patterson would be publicly humiliated.

Kehoe's comments brought up a classic dilemma in publishing scientific results. On the one hand, all science deserves to be seen and scrutinized by the world, especially when it challenges long-held assumptions. But on the other, scientists rely on the gatekeepers of prestigious journals to prevent shoddy research and half-baked conclusions from muddying their fields. With the awkwardness of Kehoe's harsh review, which wasn't exactly unbiased, Patterson's work appeared somewhere in the middle. The editorial board of the journal was split. Several researchers and business executives threatened dire consequences if Patterson's paper was published. Patterson promised equally dire consequences if it was not. The editorial board approved publication by one vote.

Patterson's work appeared in September 1965 with what would become a classic title in the fields of environmental

and public health: "Contaminated and Natural Lead Environments of Man." The paper was not publicly ridiculed, as Kehoe hoped it would be. It was excerpted in the *New York Times*
and the *Washington Post* with a conclusion so compelling—
that modern humans carry 100 times more lead than pre-
Columbian peoples and that the atmosphere holds 1,000
times more lead than in the past—that it drew an instant reaction. And an equal and opposite *re*action.

Ethyl Corporation got a tip that Patterson's paper would be
published, so it was ready to fight back. Company officials demanded a meeting at the Public Health Service in Washington to discredit Patterson and forestall any regulation. The
surgeon general at the time, William H. Stewart, felt the same
pressure that his predecessor Hugh Cumming had decades
prior, so Stewart planned a meeting for both sides to air their
opinions. More than one hundred executives in the lead industry attended, with two long tables for reporters. Unlike
the low tenor of the 1925 Public Health Service symposium,
the 1965 session was "noisy, angry, and sometimes incoherent
because of emotion," according to Harriet Hardy, the protégé
and friend of Alice Hamilton, who attended.

After the meeting, everyone went outside for a press conference. Journalists lobbed questions at the lead executives,
who stood surrounded by cameras and microphones. Kehoe
approached the scrum and started to rant about Patterson's
work and all the ways it was deficient.

But then a reporter caught Kehoe by surprise and asked
what would be a pointed and revealing question not about
Patterson but about Kehoe:

"What is your salary, who pays it, and from where does the
money come that supports your laboratory?" the reporter
wondered.

In the eyes of those who watched this scene, Kehoe sank. He paused, then sputtered, then admitted it came from the same lead industry executives he was working hard to protect.

In barely two minutes, Kehoe was effectively finished. Nothing about Kehoe's funding or his work had changed. But the country had shifted. Tetraethyl lead was no longer the exciting elixir of the future but a proven poison whose havoc on the world was becoming clear. In this light, Kehoe's admission that he was a longtime industry shill—which he had never kept secret—became newly newsworthy. In the weeks to come, every prior study of his was reinspected for errors and faulty assumptions. Compared to Patterson's technical work, which relied on cutting-edge equipment and advanced research methods, Kehoe's methods looked old and had the added weight of also being wrong.

Kehoe retired from the University of Cincinnati before the year was over. He insisted that he was planning to anyway, but it wasn't hard to see that his standing was diminished. If even half of what Patterson published was true, Kehoe would become a cautionary tale of an industry scientist profoundly co-opted and whose willful blindness had come at great cost to millions of people. In 1964, Kehoe had been so celebrated at the University of Cincinnati that a campus building was named after him. By 1966, the building was renamed.

Kehoe continued to defend himself in his retirement, but he had to contend with his own health difficulties. In 1973, he was diagnosed with an unknown illness. He had bouts of irritable and irrational behavior and was prone to outbursts and "childish ways of self assertion," according to a letter he wrote to a friend. His medical records were never made public, but descriptions of his manic behavior overlap with the clinical symptoms of moderate nonfatal lead poisoning. Kehoe died in 1992, at age ninety-nine.

Even in death, Kehoe's work lived on. The Kehoe principle, the scientific paradigm he employed and that was later named after him, shifted the burden from a company to prove its product was safe to the general public to prove that it wasn't. And with this framing, the company would win every time. Either the product *was* safe and could proceed, or if it wasn't, it would take decades for other researchers with less time, funding, and knowledge on the subject to establish sufficient evidence to warrant regulation, during which time the product could still be produced, marketed, and sold. The Kehoe principle is still used widely today.

The downslide of Ethyl gasoline began in the summer of 1966. A senator from Maine named Edmund Muskie held hearings on the lead issue. Patterson testified, and he and several other scientists provided mountains of calculations to the committee. With extensive media coverage, the public had been awakened to a poison infiltrating their food, water, and air.

The public's anxiety, measured in the bricks of letters received by congressmen and senators, in turn fueled increases in funding for scientific investigation. In 1965, the National Library of Medicine recorded twenty-one scientific studies about lead poisoning. By 1969, there were 112. Many of the new investigations looked at low-dose poisoning, an area with scant prior attention due to Kehoe's longtime insistence that low exposure to lead was perfectly normal. The new findings revealed that children absorb and retain four to five times more lead than adults. And contrary to Kehoe's longstanding poisoning threshold—80 micrograms of lead per deciliter of blood—researchers found the real number to be less than half that, 30 micrograms. After further research, the threshold was lowered to 25 and then 10. Today, the threshold for poisoning used by the Centers for Disease Control is 3.5, although it believes that no level is truly safe.

Unable to outrun such alarming findings, the Lead Industries Association saw its membership fray into a circular firing squad. Companies that made lead paint blamed companies that made lead pipes. Lobbyists for Ethyl Corporation claimed that the reason so many children had elevated blood lead was that they chewed on the lead paint on their cribs, which had nothing to do with gasoline. But arguing on technicalities won them little ground. The danger was lurking in every corner of American life.

Lurking, of course, implied that it had gotten there quietly, without anyone knowing. In many ways it had. Despite the hundreds of pages of evidence of its danger in the 1920s, much of it collected, written, and published by Alice Hamilton and Yandell Henderson, the public seemed to be genuinely surprised in the 1960s at how saturated the world was in such a well-known poison.

The logical next question was how it had been allowed to happen.

When Ethyl Corporation realized that the long reign of its cash-cow product was coming to an end, it tried to delay the inevitable.

In 1972, the new federal office called the Environmental Protection Agency held a hearing in Los Angeles. A man named Lawrence Blanchard, who was then a vice president for Ethyl Corporation—the same position once held by Thomas Midgley—had been dispatched by the company to fight any EPA efforts to regulate or phase out lead gasoline. And if he couldn't do that, he was to extend the phaseout period as long as possible to give Ethyl time to develop a replacement product. To accomplish this, Blanchard used many of the same arguments deployed by Kettering and Midgley in 1925: that tetraethyl lead had transformed American life and that without it cars would sputter, the economy would collapse, and the country would run out of oil. Blanchard called efforts to regulate Ethyl a "witch hunt."

Just as in 1925, the government officials tasked with weighing economics against public health found Blanchard's argument compelling. The new agency was also nervous about making such a sweeping and impactful decision less than two years into its existence. The EPA stalled for a year, but eventually the science and the pressure were overwhelming. The phaseout was scheduled to start in 1974. Ethyl tried to appeal the decision to a federal court of appeals, but by a decision of 5 to 4, it lost. It appealed again, but the Supreme Court refused to hear the case.

This saga was maddening to a company that had enjoyed decades of lucrative dominance. But it also had a strange twist. Because in the end, it wasn't the EPA that killed Ethyl gasoline after all.

The death blow came from a pair of engineers in New Jersey. At the same time Ethyl was fighting for its prized product against the EPA, the two inventors working for a metal fabrication company called Engelhard released a new invention for cars that would help reduce the black clouds of unburned fuel, carbon monoxide, and nitrogen oxide that came out of tailpipes. Like inventors before them, they were responding to an urgent demand from the public and also a mandate from the U.S. government, which set strict limits on air emissions in a new law called the Clean Air Act.

They called their creation a catalytic converter, and it forced the exhaust components to travel through a chamber, where they would mix with precious metals like platinum or palladium. Through simple chemistry, the converter would transform carbon monoxide into carbon dioxide, and unburned hydrocarbons into carbon dioxide and water. It was a simple and cheap way to fix a dirty problem. The only downside of the catalytic converter was that lead would clog it up.

Starting in 1975, all new cars came with catalytic converters, and when they did, gas stations began dropping leaded gasoline even faster than the EPA required them to. An invention in the 1920s to improve American automobiles was replaced by a new invention in the 1970s. A cycle of ingenuity in exactly fifty years.

It would take another ten years for the full "phasedown" (as it was gingerly called) of leaded gasoline in the United States. The lead industry requested the term to avoid harsher words like "phaseout" or "ban." But the effect was the same.

To scientists like Clair Patterson and Harriet Hardy, it appeared for the moment that the worst had passed and that a dangerous substance—and the obstinate companies behind it—had finally been tamed. No one would be poisoned anymore.

But despite all of Patterson's and Hardy's work exposing the problems with leaded gasoline, neither had looked into the sweeping effects it had on every American person and on the world more broadly. That would take time. When someone eventually did, their results were more shocking than anyone could have imagined.

# 20

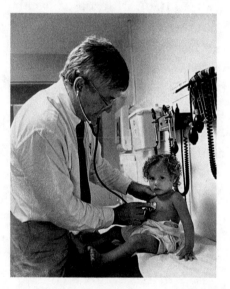

*Pediatrician Herbert Needleman inspecting a child in 1970*

## Philadelphia, Pennsylvania — 1979

Science is often about noticing patterns, of seeing one thing, seeing another, and figuring out if they're connected. This was how a young pediatrician in 1970 made the first revelation.

Herbert Needleman worked with children in South Philadelphia, a part of the inner city filled mostly with Black families. His office overlooked a playground. During the day he'd sometimes look out the window and watch the kids play. Something struck him as odd about many of them: They appeared lethargic, moved slowly, and would often stare off blankly.

The kids reminded Needleman of an old memory. Back during medical school in the 1950s, he took an internship one summer to make extra money at a DuPont plant in Deepwater, New Jersey. The work was grueling and hot. He and a group of men unloaded railcars filled with unknown chemicals and neutralized them with sulfuric acid. During breaks he would go outside and guzzle a gallon of water with the other workers. One day he noticed a pair of older workers sitting nearby. "They were gray and mute, staring off into space, not talking to anybody," he later recalled. Someone told him the men worked in an adjacent building that made a lead sludge. They called it the "House of Butterflies" because of the men's delusions.

Twenty years later, Needleman couldn't shake the image of those men as he watched the children on the playground. Could they have lead poisoning too? It seemed impossible. None of the children worked in heavy metal plants. And besides, if they were poisoned, it would have shown up in routine blood tests. All the research in the world on lead suggested that lead poisoning was binary: You either had it or you didn't. So long as a person's blood lead level was below the allowable threshold, they were not poisoned and would have no symptoms.

Yet watching the kids made him wonder if the issue was more complicated. Perhaps blood, which was often refreshed and replenished, was not the best measure of a person's long-term lead exposure. Bones would provide the best evidence of a person's lifetime absorption, but performing bone biopsies on children was out of the question. So Needleman thought of the next best thing to bones: baby teeth.

In the summer of 1971, Needleman started to collect baby teeth from first and second graders in Philadelphia. He distributed silver half dollars to local dentists—a sum higher than most kids got from the tooth fairy—and hoped teeth would come in. When they did, he found exactly what he predicted. Many of the kids had been exposed to low or even moderate amounts of lead almost their entire lives.

This was stunning. But for it to mean anything, he wanted to find out if the lead in each child somehow influenced their behavior. Needleman had the idea to enlist teachers. He asked them to fill out questionnaires about each of their students. They were simple yes/no questions: Is the child distractible? Are they disorganized? Are they hyperactive or impulsive? The teachers didn't know which students were in the study, nor did they know any student's tooth lead levels. But their responses revealed a strong correlation. As tooth lead levels went up, so did the number of bad reports. Children with the highest levels had the most trouble paying attention and were slowest to learn to read. Needleman gave IQ tests to all the students and, almost uniformly, the higher the lead in their teeth, the lower their scores.

Needleman published his findings in 1979 in the *New England Journal of Medicine*. He knew it would be controversial. But he also had a thick skin and a general conviction against injustice, however small. And the data suggested this was bigger than a small injustice. The lead industry knew what he was working on. He couldn't keep secret a public tooth collection for the purpose of lead testing. Yet he had grown desensitized to the regular threats of career assassination from lead executives and the occasional threatening calls to his house in the suburbs of Philadelphia. It helped that he and Clair Patterson at Caltech became friends in the fight. The two were so accustomed to the withering attacks that they even joked about them. Needleman wrote to Patterson in 1980, "When somebody calls me an environmental extremist, you can defend me, as I do when they call you a nut."

As time went on, Needleman's tidy correlations inspired more researchers to dig into an area almost guaranteed to produce striking discoveries. One by one, each new study zeroed in on the same source of lead responsible for the most damaging effects on public health—not paint, not pipes, but gasoline. A scientist in Baltimore tested levels in soil and found that dirt around highways contained ninety to one hundred times more lead than around rural backroads. In New Zealand, someone tested the lead levels in sheep and found elevated levels and routine health problems in animals that lived beside busy roads. Yet another scientist revealed that children fared worse than adults, and Black and poor children who lived near factories and on the routes of delivery trucks had it worst of all. An environmental health scientist at Columbia found that when children are exposed to lead dust in their early years, they become like a delayed train, arriving later and later to future developmental milestones.

As a neurotoxin, lead was well-known among doctors for the way it killed gray matter in the brain's prefrontal cortex. But in the 1980s, researchers began to probe what effect, if any, a few dead brain cells were having on children as they grew up. The prefrontal cortex is the part of the brain that regulates emotion, impulse control, and moral judgment. In places with a high consumption of leaded gasoline, Needleman and other scientists were startled to discover higher rates of high school dropouts and teen pregnancies.

What this meant for society at large wasn't clear at the time. It would take years for children born in the early 1970s, at the peak of leaded gasoline consumption, to be adults in the 1990s. And it would take more time for researchers to learn if slight delays in children's development had any noticeable effect on their lives as adults—and, by extension, whether any trends could be seen across an entire generation of similarly aged adults.

But then, in 1990, something unexpected started happening. Almost overnight, crime rates across America appeared to reach a peak, and then, rather quickly, they dropped. In the next few years the number of assaults, robberies, and murders in the country fell by as much as half. Police and prosecutors took credit for cleaning up the streets. Mayors enjoyed easy reelections.

As more years went by and crime continued to drop almost everywhere, it appeared something else might be at play. In 1997, an economist working for the Department of Housing and Urban Development named Rick Nevin noticed a strange coincidence in trends. The line graph showing the consumption of leaded gasoline over time looked like a camel's hump: It rose in the 1940s and 1950s, peaked in the 1970s, and then declined. He realized that the graph showing violent crime was the exact same camel's hump but twenty years delayed. Crime rose in the 1960s and 1970s and peaked in the 1990s, then declined. He sought more detailed data to more closely inspect the graphs of lead emissions and crime rates to see if the correlation was as clear as it seemed. And it was. The lag between the two graphs was exactly twenty-three years. Children who ingested high levels of lead in their first two years appeared to be more likely to become criminals by the time they were twenty-five—the exact age when their brains were supposed to be fully developed if they had not been partially damaged by lead.

This was a compelling coincidence, but it wasn't proof. Many other social and economic factors could have contributed to the decline in crime over such a long period. The 1990s saw increases in police officers and the end of the crack epidemic. The legalization of abortion twenty years earlier suggested fewer unwanted babies in the 1970s who might

# THE LEAD-CRIME HYPOTHESIS

Historical correlation of lead consumption and violent homicides separated by 21 years

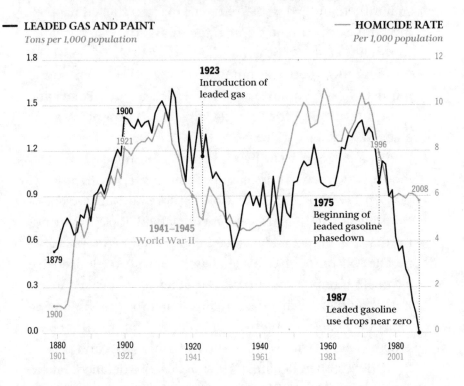

**— LEADED GAS AND PAINT**
*Tons per 1,000 population*

**— HOMICIDE RATE**
*Per 1,000 population*

**1923**
Introduction of
leaded gas

**1975**
Beginning of
leaded gasoline
phasedown

**1941–1945**
World War II

**1987**
Leaded gasoline
use drops near zero

Homicide data from the CDC, lead data from the USGS. First published by Rick Nevin,
"How Lead Exposure Relates to Temporal Changes in IQ, Violent Crime, and Unwed
Pregnancy," *Environmental Research* 83, no. 1 (May 2000).

grow into dejected adults. But then a graduate student at
Harvard thought of a way to test the strength of Nevin's correlation. The student, named Jessica Wolpaw Reyes, wanted
to localize the data. If reductions in leaded gasoline truly fueled a national reduction in crime, then states that phased out
leaded gas quickly would see quick drops in crime, and states
that phased it out more slowly would see slower drops. Reyes
collected data from every state that would give it to her; to
her delight and horror, the correlations lined up perfectly.
Additional studies narrowed in, comparing one neighbor-

hood's car pollution to the next. Others looked internationally, sizing up one country's gasoline and crime rates versus another's. In every instance, on every chart, the rise and fall of leaded gasoline and the rise and fall of crime almost perfectly correlated.

Today, this is known as the lead-crime hypothesis in criminology. It remains a hypothesis because it will forever be impossible to prove and also, perhaps, because public policy officials and law enforcement experts like to believe that billions spent on crime reduction and social justice programs for decades have worked, rather than credit a factor as dull as car fuel. But by the same token, the links between leaded gasoline and crime have never been debunked. And every year, economist Rick Nevin still gets excited when new crime data comes out and he sees his predictions confirmed. "The trends have continued in an astonishing way," he told me.

Meanwhile, research continues on more detailed questions, like how little lead it takes to kill brain cells; how losses in emotional regulation and impulse control have led to higher rates of ADHD; and how the damage is often worse in boys than girls. One especially grim area of research investigates the lead dust still embedded in soils and subsoils from gasoline burned decades ago and the ways rainstorms and construction projects return it to groundwater and air. Studies have shown that in major cities around the world, low levels of tetraethyl lead still float in the air and probably always will.

In 2002, the United Nations Environment Programme launched an initiative to eliminate leaded gasoline worldwide. More than forty countries still used it at the time, including almost every country in Africa. One by one, countries slowly dropped the fuel with the help of private grants and international aid to remake supply lines and replace old equipment. In 2021, exactly one hundred years after Thomas

Midgley first fed tetraethyl lead into a test engine in Dayton, Algeria was the last country to phase it out.

Ethyl Corporation found a way to hang on. Using its fifty years of profits from tetraethyl lead, Ethyl began diversifying its portfolio in the 1970s to include oil, coal, aluminum, paper, and plastic. One new division made pain medication. Another sold Ethyl life insurance.

The company's chairman at the time, Floyd Gottwald Jr., thought about changing Ethyl's name to sever any association with lead. But the brand was embedded in Americans' minds after decades of saying "Fill 'er up with Ethyl." And besides, the name didn't mention lead at all, just as Charles Kettering intended when he coined it in 1923.

Ethyl Corporation avoided liability for profiting off a product its executives knew was poisonous for fifty years by investing in softer forms of reputational management. It made charitable donations to museums and arts organizations. It funded credible scientists with large grants. Even as it phased out tetraethyl lead in the United States, it expanded its business overseas tenfold between 1964 and 1981. This expansion, along with its investments in other chemical products, kept its stock rising every year, often by double digits. That was enough to repress any sense of corporate shame. In the case of unfavorable press or protests outside its Virginia headquarters, Ethyl could use its flush reserves to highlight all the good it was funding and all the smart people who supported it. In 1980, Ethyl invited the chairman of the National Geographic Society, Gilbert Grosvenor, to be on its board of directors. Grosvenor didn't know what the company did, but he remembered the name from when he was a kid, so he accepted. "I don't recall very much about the experience," Grosvenor said recently, "but the pay was generous."

Meanwhile, Charles Kettering, who founded Ethyl Corporation under General Motors in 1923, also found ways to avoid any reputational damage as the dangers of his prized product were unmasked. In 1927, as sales of Ethyl gasoline started to reach new heights, Kettering donated 10,000 shares of GM stock worth $2 million to start a research foundation in Ohio named after him. His goal was to bring together medical experts to, as he saw it, "carry out scientific research for the benefit of humanity."

This gave Kettering, who was then fifty-one, the position of elder statesman in science and engineering, which yielded additional benefits. Kettering was able to selectively fund younger scientists and direct the course of their research. He was a sought-after speaker and was regularly encouraged to run for public office. In January 1933, Kettering appeared on the cover of *Time* magazine as a driving force behind American progress. In 1940, President Franklin D. Roosevelt appointed him to head the National Inventors Council.

With every year, Kettering grew richer and richer. His decades of generous pay and royalties compounded in the rising stock market after World War II. His list of awards also grew. He received medals for science, engineering, and electrical arts, along with thirty honorary degrees. Every so often there would be legal squabbles for Ethyl to deal with, and occasionally Kettering would have to appear in court. The biggest case, in 1940, was an antitrust lawsuit from the U.S. government that alleged Ethyl was using its market dominance to exclude other companies that wanted to make and sell leaded gasoline. But none of the lawsuits posed any meaningful danger to Ethyl, much less Kettering himself.

Kettering died in 1958. His lifetime of generous donations and business connections bought him a sterling legacy. A suburb of Dayton named itself Kettering in 1955. At least four universities—the University of Cincinnati, the University of

Dayton, Ashland University, and Oberlin College—named science and medical centers after him. In 1998, the General Motors Institute, a private Michigan college devoted to automotive research, wanted to rebrand itself as a center for broader science and invention. Its board of trustees chose the new name Kettering University.

One act of legacy preservation, however, stood head and shoulders above the rest. In 1945, Kettering's former boss, General Motors CEO Alfred P. Sloan, donated $4 million to a small medical center on the east side of Manhattan. Sloan convinced his friend and former colleague Kettering, who was then sixty-nine, to donate more money, plus his time, and oversee a research center adjacent to the hospital. Their novel idea was to apply industrial research methods to medicine, meaning a rigorous trial-and-error approach to curing illnesses.

In 1980, the Sloan Kettering Institute, as it was named, was combined with Memorial Hospital. Today it is known as the Memorial Sloan Kettering Cancer Center. It treats four hundred types of cancer and is routinely ranked one of the top cancer hospitals in the world.

It is a bitter truth that the leaded gasoline that Kettering, Midgley, and Sloan created and sold for decades is linked to cancer. The United Nations' International Agency for Research on Cancer now considers organic lead from leaded gasoline that remains in contaminated air, dust, soil, and water to be a carcinogen in mammals.

Meanwhile, doctors at Sloan Kettering and other cancer centers continue to treat rising rates of cancers worldwide, with especially high numbers in adults younger than fifty. According to a 2022 study from the World Health Organization, each generation now appears to have a higher risk of developing cancer than the generation before it.

Research is ongoing as to its cause.

After Alice Hamilton died in 1970, the world changed again. She was lucky to live to 101, especially after a lifetime of working with toxins that were known to kill people even at low exposure. But she was also deprived of seeing what happened next. First came the sprawling 1970s era of federal and state laws that protected workers' rights, clean air, and drinking water. Then came the downfall of tetraethyl lead that she had so strongly fought for. And then, with each passing year, the public seemed to awaken to more of the toxins that she spent her life studying and warning about. She was vindicated on the dangers of low-dose poisoning from mercury, radium, asbestos, and carbon monoxide.

The more Hamilton was proven right, however, the further she seemed to fade from public view. She became less remembered, less quoted, and hardly credited at all when new bombshell studies on toxins appeared in the 1970s and 1980s that were built upon her work. In 1985, when the EPA was overseeing the phasedown of leaded gasoline and coming to terms with the effects of such a disastrous blunder, many senior officials at the EPA—an agency filled with environmental experts—were baffled to learn that someone had predicted in 1921 the destruction tetraethyl lead would cause almost the moment it was mixed with gasoline.

Hamilton's diminished legacy was partly her own doing. In her lifetime she did not command attention, seek out fame, or try to enlarge her profile. She did not make large endowments that would adorn buildings with her name or fund future students with scholarships. By virtue of not marrying, she did not have extended family or children who might have promoted her legacy or safeguarded her records. History holds in high regard people who reach new frontiers in medicine, like Clara Barton and Elizabeth Blackwell, and those responsible for saving thousands of lives, like Alan Turing and Oskar Schindler. But legacies generally go to people who want them and plan for them. Hamilton took little time to consider hers and ended up with a proportionally small one.

And yet the tendrils of Alice Hamilton's work reach deep into every aspect of modern life. She laid the foundation for occupational health and safety standards that protect millions of workers worldwide. Her research directly informed new regulations on toxins and disease in dozens of countries, as well as the creation of new government agencies to enforce them. She broke gender barriers both at Harvard and in the broader sciences that opened stodgy male-dominated fields to women who would make their own discoveries. And her approach to social justice—combining evidence-based research, interdisciplinary collaboration, and community engagement—remains the blueprint for nearly all public health and policy fights today, from the smallest neighborhood disputes to the global battles over pollution, natural resources, and climate change.

In this way, Hamilton belongs in the hall of famous environmentalists, alongside Teddy Roosevelt and John Muir, but with a more complicated view: The environment worth protecting is not just in national parks and beautiful places but everywhere, even underground in a mercury mine. Clean air, water, and soil are worth having not just because they sound nice but because the opposite can be deadly. These were once controversial ideas, but after a century of Alice Hamilton and other environmentalists repeating them with data to prove it, they've become almost obvious and cliché.

In 1995, Hamilton was honored on a U.S. postage stamp in a series of "Great Americans." That same year, it was becoming clear that the work she started in industrial toxicology wasn't finished and in many ways was just beginning. Leaps in technology, especially in digital systems, had spurred another revolution in industry that brought thousands of new chemical materials. Almost daily, companies were developing new synthetic chemicals to make computers smaller, to make cars more powerful, and to make food grow faster, last longer, and taste better. Revolutionary compounds made cheaper textiles, nonstick cookware, and a ballooning number of plastic products.

All these new materials required factories to make them and workers to run giant machines. A new organization called the Association of Professional Industrial Hygienists started offering credentials for people who wanted to do the work Hamilton had done almost a century prior. And when they got certified, a growing army of industrial hygienists and health and safety engineers fanned out into warehouses and manufacturing plants to do things like test indoor air quality, run safety workshops, and make sure nobody got their arm sliced off. As more people entered the field, another industry group started offering an annual "Alice Hamilton Award" to women devoted to worker health, social reform, and scientific truth.

This was how Alice Hamilton saw success: not in winning every battle but by bending the arc of history more toward justice. Several years before she died, Hamilton looked back at the landscape of her life and surveyed the hills and valleys, the wins and losses. "I wouldn't change my life a bit," she told a friend. "For me the satisfaction is that things are better now, and I had some part in it."

In 2000, the city of Fort Wayne, Indiana, erected a statue of Hamilton, along with statues of her sister Edith and her cousin Agnes, in a park beside the Saint Marys River. Alice's statue is tucked away behind trees, visible only to those who know it's there. Sometimes on spring mornings, a sparrow will land on her head and sing while the sun comes up.

# Epilogue

On the west side of Richmond, Virginia, in the neighborhood of Gambles Hill, stands a painted white building with four Ionic columns that so closely resembles the White House that visitors sometimes mistake it for the Virginia state capitol. When it was built in 1958, the architects looked at Colonial Williamsburg for inspiration. They wanted to give the building a sense of history and prestige befitting the headquarters of a dominant U.S. company.

The headquarters of Ethyl Corporation still stands today. Its faded bricks and tarnished colonnade show signs of time passing. Its staff has thinned over the years, from thousands of employees to a few hundred today. One hundred years after its founding, the company created by Charles Kettering and Alfred Sloan in 1924 has changed owners several times. It's now a subsidiary of a chemical company called NewMarket Corporation, also based in Richmond, which makes a broad portfolio of petroleum additives.

Both NewMarket and Ethyl have never fully confronted their role in what the UN Environment Programme has called one of the biggest environmental catastrophes in human history. Today the chairman of NewMarket is Thomas

E. "Teddy" Gottwald, the grandson of former Ethyl president Floyd Gottwald. The Gottwald family is regularly ranked as one of the wealthiest families in America and is often recognized for its charitable donations. But it has never publicly acknowledged the effects of its foundational product. Over the course of three years, I repeatedly asked Gottwald and his leadership team to make available his company's corporate archives, sit for an interview, or respond to questions in writing about Ethyl Corporation's history of producing and promoting leaded gasoline. Gottwald declined via NewMarket's general counsel, Bryce Jewett, who told me that the company considered the matter closed. "The history of tetraethyl lead, Ethyl Corporation, and its parents in the 1920s and 1930s is well documented," Jewett said. "We do not have anything to add."

Ethyl's archives are thought to contain many more details about this story, including medical reports about employees who got sick during tests of tetraethyl lead and who in some cases spent decades after in psychiatric institutions. Prior court cases have alleged the existence of correspondence in Ethyl's possession between Ethyl's leaders acknowledging the superior alternatives to tetraethyl lead and also the naked profit calculations for making and disseminating a substance the company's top officials knew was poisonous. In various cases, Ethyl denied these accusations and asked judges to keep its internal documents under seal.

The richest series of documents is known as the "Lead Diary," a term coined by Charles Kettering and T. A. Boyd in the 1940s, when they recalled their memories of events in the 1920s. The legal statutes of limitations for liability have long passed, as have all the people involved. But according to several sources with knowledge of these documents, the Lead Diary contains admissions from former top officials that would still be embarrassing to several companies today, including NewMarket, Ethyl, General Motors, DuPont, and Exxon Mobil (formerly known as Standard Oil of New Jersey). I sought on-the-record interviews with senior leaders at these companies as well. They also declined. Several people at these companies reached out to me privately, offering to talk if they could remain anonymous. One of them reflected in a long phone call about, in his words, "the philosophical nature of liability." By this he meant the legal structures that enable a company to claim credit when things go right and usually evade blame when they go wrong. He admitted—in a way that struck him as almost laughably obvious—that it had become common among large companies to follow the defensive strategy deployed by Sloan, Kettering, and Midgley a century ago: When faced with crisis, liability, or embarrassment, deny wrongdoing, deflect accusations, and discredit opponents—and if all else fails, delay regulation as long as possible. There's a reason it stuck. With enough time and money, it usually works.

Amid these realities, it's easy to see more examples of this story in today's world, of people and companies that reap the rewards of their advances but socialize the costs (and then deploy elaborate disinformation campaigns to keep the train rolling). Obvious candidates include the makers of fossil fuels, synthetic pesticides, and of course cigarettes. But there are many more we can't see, experiments happening around us in pursuit of making our lives easier, cheaper, cleaner, or tastier. When it comes to the building materials in our houses, the compounds in cleaning supplies, or the synthetic additives in our food, we mostly assume that someone confirmed everything's fine. You kind of have to. To live your life afraid of everything is corrosive in its own way.

But plenty of evidence proves that there's less supervision of these products and companies than we'd like to think. Industries that promoted opioids, sold subprime mortgages, and emitted billions of tons of greenhouse gases have faced consequences only when it's too late—and sometimes not at all. Researchers have found elevated levels of heavy metals in imported products like vitamin supplements, jewelry, and children's toys. Every American is promised clean drinking water, unless you live in Flint, Michigan, or Jackson, Mississippi.

Over the course of writing this book, my wife and I would sometimes discuss at our dinner table what could be the poisons around us today, the ones we'll discover in fifty years were devastating blunders. (People love us at dinner parties.) It's hard to know, but I'm sure that for all of them, there are Alice Hamiltons at this moment trying to warn us. There are copious plastics and silicones around our house that researchers have found to be . . . probably safe. I suspect we'll have bigger regrets about—once again—a type of industrial chemical that has made life better but remains only lightly understood. If you haven't heard of perfluorinated and polyfluorinated alkyl substances, also known as PFAs (and also known as "forever chemicals" for the way their molecular structure never breaks down), you will eventually. They're in almost every consumer product, from food packaging to shampoo to dental floss, and have already been linked to high blood pressure and certain types of cancer. When their full impact is known, it will probably take decades, just like with leaded gasoline, to cast them off.

But I've come to find a more optimistic way to view this story and others like it unfolding around us today. If one person, one group, or one company can cause so much harm for their own personal gain, then the opposite is also true. A single person who's prepared and perseverant can create considerable progress for the good of others.

Alice Hamilton spent her life determined to prove this was possible. She lived in a world of limitless threats and had endless opportunities to despair about humanity poisoning itself. And yet she didn't. She was a quiet yet determined activist who spent her life demonstrating the compounding power of diligence and repetition. It took one hundred years, from 1921 to 2021, for leaded gasoline to run its course. But in another sense, it took *only* one hundred years. The same awaits those who fight for healthy communities, clean natural resources, and a planet in balance. The tide often changes too slowly, but it does change.

# Acknowledgments

After writing this book, my biggest thanks go to those who make modern life possible: the unsung plant workers, laborers, technicians, drivers, and organizers who tend to only get noticed when they screw up, and they don't screw up much.

For the third time, my intrepid agent Lauren Sharp believed in me and this project, and, also for the third time, she captured the imagination of Dutton about what it could be. My editors Brent Howard, Grace Layer, and Emi Ikkanda strengthened this book at every stage, as did Evelyn Duffy and Ben Gambuzza.

My research stands on the shoulders of prior scholars who have devoted their careers to the subjects in this book. I could not have written this without Bill Kovarik, with his encyclopedic recall of Ethyl history, and Barbara Sicherman, with her almost spiritual ability to decipher Alice Hamilton's handwriting. Special thanks also to Jamie Kitman, Gerald Markowitz, David Rosner, Stuart Leslie, Angela Nugent Young, and Alan Loeb.

Behind every nonfiction author is an army of archivists and librarians, all of them devoted to keeping records accessible, and many of them extremely nice people too. Connie

Carter of the Library of Congress has long been my friend and yoda, excitable at the slightest whiff of a research quest. The LOC's Joanna Colclough, Kelly Abell, Amber Paranick, and Natalie Burclaff helped me locate, with patience and kindness, endless records, including rare newspaper articles and scientific journals. Sarah Hutcheon and Jennifer Fauxsmith at Harvard's Schlesinger Library were my guides through Alice Hamilton's sprawling collection of records, as was Cathy Hajo, who oversees the Jane Addams Papers Project at Ramapo College. Additional thanks to Linda Gross at Hagley Library, Ginny Kilander and Nora Plant at the University of Wyoming, Greg Taormina at the National Archives in College Park, Glen Longacre at the National Archives in Chicago, Adam Stevenson at the Small Library at the University of Virginia, Gabrielle Barr at the National Institutes of Health, and Eve Brant at *Harper's Magazine*. Much of this book relied on the collection of corporate documents obtained through donation and legal discovery and housed (for free) at ToxicDocs.org. Toxic Docs is run by Merlin Chowkwanyun and funded by several public-spirited foundations.

Speaking of legal discovery, several anonymous sources gave me internal corporate documents detailing decision-making at Ethyl Corporation, DuPont, and Standard Oil, many of which have never been made public. A handful of people at these companies also risked their careers and pensions to talk to me. I promised not to name them, but I thank them for their courage. The ones I can name and thank are Ron Richardson, formerly of Angelos Law, and Collette McDonough of the Kettering Foundation.

This book was supported by the Smithsonian National Museum of American History and the Lemelson Center for the Study of Invention and Innovation, which awarded me the Arthur Molella Distinguished Fellowship on the history of invention. I was first excited about visiting the Smithsonian's voluminous collections, but I quickly learned the Smithsonian's even greater asset is its people. My endless thanks to Eric Hintz and Alison Oswald for the friendship and encouragement, and to Jeffrey Stine and Trina Brown for the kindness and help navigating the sprawling Smithsonian archives.

Several research assistants know my idiosyncrasies and research quirks (*always more weather details!*). I'm grateful for the digging and diligence of Patty Kessler in Wyoming, Scott Burgh in Chicago, and Kristen Smith and Erin Aslami in Boston, and for the bibliographic brilliance of Morgan Miller in Maryland. Matt Twombly both meticulously researched and masterfully illustrated some of the complex subjects in this book.

My writers' group of Liz Flock and Lance Richardson provided their usual gut checks and encouragement. It feels like an afterthought to say it so late, but the spark for this book came from Liz. Everyone needs an industrial hygienist in their life; mine is Rachel Zisook, who is surely a great friend too. Shaena Montanari, George Zaidan, and Justin and Kristi Greco provided scientific consults to make sure I didn't sound like an idiot. (Any remaining errors are mine, not theirs.) Jeff Bartholet and Bill Press critiqued drafts and made this book better.

As in chemistry, my nucleus is my everything. My endlessly supportive wife, Alanna, is the fuel that makes my work and world possible. Charlie is my loyal lap sitter. And Micah and Jonah are our future, who remind me every afternoon when it's time to quit writing and build some towers.

# Notes

This is a work of nonfiction about real people and real events. It is also a book about a long-ago controversy involving many people, all of whom have died. Many of the characters in this story often had different and sometimes conflicting recollections of the same events. In general, I prioritized contemporaneous first-person primary sources. When that wasn't possible, I used details acquired through second-hand recollections, always considering the motives of the people doing the recalling. In such a sprawling story, I also relied on the secondary material of prior scholars who reviewed archives that I was unable to visit, are no longer accessible, or may have been destroyed. In limited cases, I made logical assumptions about people and events based on corroboration from independent sources.

## Chapter 1

4    **Harvard University had expanded:** "The History of Harvard University," Harvard University, last modified September 22, 2023, https://www.harvard.edu/about/history/.

4    **David Edsall, who, outside of:** "The Founders & Deans of HSPH," Harvard T. H. Chan News, last modified June 30, 2016, https://www.hsph.harvard.edu/news/magazine/centennial-founders-deans/.

4    **beige turtleneck with a brown blazer:** Letter from Alice Hamilton to Agnes Hamilton, October 1920, Schlesinger Library, Harvard University.

5    **Montgomery Hamilton, pushed Alice and:** Barbara Sicherman and Alice Hamilton, *Alice Hamilton: A Life in Letters* (Cambridge, MA: Harvard University Press, 1984), 17–18.

5     **"whose influence would shine":** Nancy Davis and Barbara
      Donahue, *Miss Porter's School: A History* (New Haven, CT:
      Northeast Graphics, 1992), 7–8.

5     **"There may be dancing":** Davis and Donahue, *Miss Porter's School.*

6     **one of the store owners:** "Will Study Health of Store Employes
      [*sic*]," *Boston Globe,* January 15, 1919, 3.

8     **Sinclair wrote of men who:** Upton Sinclair, *The Jungle* (New York:
      Jungle Publishing, 1906), 117.

8     **In 1900, 2 million U.S. children:** Walter I. Trattner, *Crusade for the
      Children: A History of the National Child Labor Committee and Child
      Labor Reform in America* (Chicago: Quadrangle Books, 1970), 71.

9     **"Shoe-leather epidemiology":** Sicherman and Hamilton, *Alice
      Hamilton,* 166.

9     **one-to-one relationship:** Sicherman and Hamilton, *Alice
      Hamilton,* 180.

9     **"make the world better":** Sicherman and Hamilton, *Alice
      Hamilton,* 407.

9     **"I am writing":** Letter from David Edsall to Alice Hamilton,
      December 27, 1918, Schlesinger Library, Harvard University.

10    **a survey for the Department of Labor:** Ian J. Lawson, "Alice
      Hamilton, 1869–1970, 'The Mother of Occupational Medicine,'"
      *Occupational Medicine* 68, no. 4 (June 2018): 224–25, https://doi
      .org/10.1093/occmed/kqy002.

11    **How about half-time?:** Letter from David Edsall to Alice Hamilton,
      January 24, 1919, Schlesinger Library, Harvard University.

11    **$2,000 a year, a sum:** "Report Increase in Positions Secured by
      Appointment Office," *Harvard Crimson,* April 13, 1923.

11    **"Isn't this wonderful?":** Letter from Alice Hamilton to Edith
      Hamilton, late January 1919, Schlesinger Library, Harvard
      University.

12    **Edsall was enthusiastic about her:** Sicherman and Hamilton,
      *Alice Hamilton,* 210.

12    **During the closed-door meeting:** Alice Hamilton, *Exploring the
      Dangerous Trades* (Ann Arbor: University of Michigan Press, 1947),
      252–53.

12    **On her first day:** Hamilton, *Exploring the Dangerous Trades,* 253.

13    **"I'm not the first woman":** "Miss Alice Hamilton, Harvard
      Professor," *Boston Globe,* March 23, 1919, 65.

13    **"The Last Citadel" . . . "break down the sex barrier":** Sicherman
      and Hamilton, *Alice Hamilton,* 237.

**13**     **the Nineteenth Amendment passed:** "Women Have the Vote!,"
Headlines & Heroes, Library of Congress, last modified November
3, 2020, https://blogs.loc.gov/headlinesandheroes/2020/11/women
-have-the-vote/.

## Chapter 2

**15**     **"a greeting from":** "Weather," *Dayton Daily News*, Weather
Circular, March 1916. The term "Eskimo" may be considered
derogatory and has a complex and frequently racist history;
however, I have preserved the original language in the 1916
article here.

**15**     **Midgley had wanted to be:** Thomas Midgley, *From the Periodic
Table to Production: The Biography of Thomas Midgley, Jr., the
Inventor of Ethyl Gasoline and Freon Refrigerants* (Corona, CA:
Stargazer Publishing, 2001), 1–15.

**16**     **he landed on the sap:** Midgley, *From the Periodic Table*, 4.

**17**     **The world's first automobile:** Steven Parissien, *The Life of the
Automobile: A New History of the Motor Car* (London: Atlantic Books,
2013), 3.

**17**     **Enormous piles of urine-soaked dung:** Águeda García de
Durango, "New York, Manure and Stairs: When Horses Were the
Cities' Nightmares," *Smart Water Magazine*, June 9, 2019, https://
smartwatermagazine.com/blogs/agueda-garcia
-de-durango/new-york-manure-and-stairs-when-horses-were
-cities-nightmares.

**17**     **"See America First":** Marguerite Shaffer, *See America First: Tourism
and National Identity, 1880–1940* (Washington, D.C.: Smithsonian,
2013).

**18**     **He enrolled in Cornell:** Charles Franklin Kettering, *Biographical
Memoir of Thomas Midgley, Jr., 1889–1944* (Washington, D.C.:
National Academy of Sciences, 1947), 363.

**18**     **a copy of the periodic table:** Sharon Bertsch McGrayne,
*Prometheans in the Lab: Chemistry and the Making of the Modern
World* (New York: McGraw-Hill, 2001), 97.

**18**     **one of the balls would float:** Midgley, *From the Periodic Table*, 7.

**19**     **Kettering's groundbreaking invention:** Stuart W. Leslie, *Boss
Kettering* (New York: Columbia University Press, 1983), 44–46.

**20**     **He wanted good people working:** Rosamond McPherson Young,
*Boss Ket: A Life of Charles F. Kettering* (New York: Longmans, Green,
1961).

**20**     **"It's important work":** Thomas Midgley, unpublished typescript,
Midgley family archives, 2022. Reviewed by the author.

2.1   **"Midge," as he called him:** Young, *Boss Ket,* 135.

2.1   **He found a list of history's:** Thomas Midgley, "Presidential Address," *Chemical & Engineering News Archive* 22, no. 19 (1944): 1646–49.

2.1   **He wore a brown suit:** Photograph, 1917, Midgley family collection. Reviewed by the author.

2.1   **"Hey, bud, the architects":** McGrayne, *Prometheans in the Lab*, 81.

2.2   **the railcar network would help:** Fred Bartenstein, "TimeLine: The Planning of Dayton," City of Dayton, updated June 2004, https:// www.daytonohio.gov/DocumentCenter/View/290/Time-Line -Dayton-PDF.

2.2   **Midgley was a family man:** McGrayne, *Prometheans in the Lab,* 80.

2.3   **Midgley's expected salary of $1,050:** Gilson Willets, *Workers of the Nation: An Encyclopedia of the Occupations of the American People and a Record of Business, Professional and Industrial Achievement at the Beginning of the Twentieth Century* (New York: P. F. Collier and Son, 1903), 1047.

2.3   **in 1913, the Miami River had flooded:** Bartenstein, "TimeLine: The Planning of Dayton."

2.3   **One man was testing new spark plugs:** Ethyl Corporation Public Relations Department, *History of Ethyl Corporation*, July 1, 1951.

2.4   **the hydrometer Kettering had asked for:** McGrayne, *Prometheans in the Lab,* 11.

2.4   **"new and useful improvement":** T. Midgley Jr., "Patent 1,432,773: Hydrometer," filed July 30, 1917, granted October 24, 1922.

2.4   **"What do you want":** Midgley, *From the Periodic Table,* 11.

## Chapter 3

2.5   **Every morning began at exactly six thirty:** Mary Lynn McCree Bryan and Allen Freeman Davis, *100 Years at Hull-House* (Bloomington: Indiana University Press, 1990), 46.

2.6   **After Addams graduated from the new Smith:** Jane Addams, *Twenty Years at Hull-House* (Champaign: University of Illinois Press, 1990), 87–90.

2.6   **The wooden shanty huts that housed:** Madeleine Parker Grant, *Alice Hamilton: Pioneer Doctor in Industrial Medicine* (London: Abelard-Schuman, 1967), 56–59.

2.7   **"A bridge between the classes":** Alice Hamilton, *Exploring the Dangerous Trades* (Ann Arbor: University of Michigan Press, 1947), 59.

28    **"We have in America a fast-growing":** Jane Addams, "The Subjective Necessity for Social Settlements," speech delivered in Plymouth, Massachusetts, 1892.

29    **Hamilton said that she wanted:** Hamilton, *Dangerous Trades*, 54.

29    **"I was resolved that in some way":** Hamilton, *Dangerous Trades*, 55.

29    **"It was not in a spirit of youthful":** "A Woman of Ninety Looks at Her World," *Atlantic Monthly*, September 1961.

30    **three dollars per week:** Addams, *Twenty Years at Hull-House*, 295.

30    **"These people seldom sing":** Abraham Bisno, *Abraham Bisno, Union Pioneer: An Autobiographical Account of Bisno's Early Life and the Beginnings of Unionism in the Women's Garment Industry* (Madison: University of Wisconsin Press, 1967), 169.

31    **"How old are you?":** Letter from Alice Hamilton to Agnes Hamilton, October 13, 1897, Schlesinger Library, Harvard University.

32    **"Miss Addams still rattles me":** Bryan and Davis, *100 Years at Hull-House*, 58.

32    **three weekly shifts in the well-baby clinic:** Hamilton, *Dangerous Trades*, 69.

33    **"I gave him the breast":** Hamilton, *Dangerous Trades*, 69–70.

34    **two Italian workmen were sitting:** Hamilton, *Dangerous Trades*, 76–77.

35    **They all worked at a nearby stockyard:** Catherine Weber, "Alice Hamilton, M.D., Crusader Against Death on the Job," 1995, Linda Lear Center for Special Collections and Archives, Connecticut College, New London, CT.

## Chapter 4

37    **The Egyptian appreciation for lead:** Herbert Needleman, "History of Lead Poisoning in the World," undated typescript, Center for Biological Diversity, Tucson, AZ.

38    **the Roman Empire, which considered lead:** Christian Warren, *Brush with Death: A Social History of Lead Poisoning* (Baltimore: Johns Hopkins University Press, 2000), 19–20.

38    **as much as 250 milligrams:** Needleman, "History of Lead Poisoning."

39    **Settlers in Virginia started mining:** "Agriculture and Industry, Virginia Main Street Communities," U.S. National Park Service, accessed February 15, 2023, https://www.nps.gov/nr/travel/vamainstreet/agriculture.htm.

40    **"the useful metal":** Warren, *Brush with Death*, 2.

**40**    **"Living in a working-class quarter":** Alice Hamilton, *Exploring the Dangerous Trades* (Ann Arbor: University of Michigan Press, 1947), 114.

**41**    **famous dignitaries and even celebrities:** Mary Lynn McCree Bryan and Allen Freeman Davis, *100 Years at Hull-House* (Bloomington: Indiana University Press, 1990), 66, 233.

**42**    **Those with science or medical experience:** Bryan and Davis, *100 Years at Hull-House*, 65.

**42**    **The first incident occurred in 1906:** Alice Hamilton, "A Woman of Ninety Looks at Her World," *Atlantic Monthly,* September 1961, https://www.theatlantic.com/magazine/archive/1961/09 /a-woman-of-ninety-looks-at-her-world/658258/.

**42**    **a young journalist named William Hard:** William Hard, *Injured in the Course of Duty* (New York: Ridgway, 1910).

**44**    **"Every year the stream":** Hard, *Injured*, 68.

**44**    **Someone gave her a book:** Hamilton, *Dangerous Trades*, 115.

**44**    **"the dangers to life":** Thomas Oliver, *Dangerous Trades: The Historical, Social, and Legal Aspects of Industrial Occupations as Affecting Health, by a Number of Experts* (London: John Murray, 1902), vii.

**45**    **"In those countries":** Hamilton, *Dangerous Trades*, 115.

**45**    **Henderson had recently returned from Germany:** Hamilton, *Dangerous Trades*, 118.

**45**    **Henderson spent months inspecting:** Charles R. Henderson, "The Logic of Social Insurance," *Annals of the American Academy of Political and Social Science* 33, no. 2 (1909): 41–53.

**46**    **"thoroughly investigat[ing] causes":** *Report of Commission on Occupational Diseases to His Excellency Governor Charles S. Deneen,* January 1911 (United States: Warner Printing Company, 1911), 5.

**46**    **$15,000 and exactly nine:** *Report of Commission on Occupational Diseases*, 6.

**46**    **"We were staggered by the complexity":** Hamilton, *Dangerous Trades*, 119.

**47**    **"premature senility":** Barbara Sicherman and Alice Hamilton, *Alice Hamilton: A Life in Letters* (Cambridge, MA: Harvard University Press, 1984), 157.

**47**    **"trying to make one's way":** Stephanie Sammartino McPherson, *The Workers' Detective: A Story About Dr. Alice Hamilton* (Minneapolis: Lerner Publishing Group, 2011), 37.

**48**    **After a month she had a list:** *Report of Commission on Occupational Diseases.*

48 **"A model factory":** Hamilton, *Dangerous Trades*, 9–11.

49 **One man told her about:** *Report of Commission on Occupational Diseases*, 24.

50 **"moral duty of every civilized state":** *Report of Commission on Occupational Diseases*, 5.

50 **After Hamilton showed him proof:** Hamilton, *Dangerous Trades*, 10–11.

51 **the Illinois legislature passed a law:** "First Workers' Compensation Law in Illinois (1911)," Office of the Illinois Secretary of State, accessed March 18, 2023, https://www.ilsos.gov /departments/archives/online_exhibits/100_documents/1911 -worker-comp-law.html.

52 **a garment factory in New York City:** David Von Drehle, *Triangle: The Fire That Changed America* (New York: Atlantic Monthly Press, 2003), 237.

## Chapter 5

56 **3 million and counting in 1916:** Stacy C. Davis, Susan W. Diegel, and Robert G. Boundy, *Transportation Energy Data Book*, Edition 33, ORNL-6990 (Oak Ridge, TN: Oak Ridge National Laboratory, 2014), Tables 3.5 and 3.6.

56 **The sludgy black gunk:** Jamie Kitman, "A Brief History of Gasoline: A Century and a Half of Lies," Jalopnik, May 21, 2021, https://jalopnik.com/a-brief-history-of-gasoline-the-lie-of-leaded -gas-1846790331.

56 **Abraham Gesner started experimenting:** Kendall Beaton, "Dr. Gesner's Kerosene: The Start of American Oil Refining," *Business History Review* 29, no. 1 (1955): 28–53, https://doi .org/10.2307/3111597.

56 **considered himself an early environmentalist:** Charles Singer, *A History of Technology* (Oxford: Clarendon Press, 1958), 105.

56 **Uncorking the pressure:** Kitman, "A Brief History of Gasoline."

57 **"reliable standards of quality":** Leslie E. Grayson, *Who and How in Planning for Large Companies: Generalizations from the Experiences of Oil Companies* (Basingstoke, UK: Palgrave Macmillan, 1987), 213.

57 **In one demonstration in 1906:** "January 26, 1906: Fred Marriott Lets Off Some Steam," *Wired*, January 26, 2012, https://www.wired .com/2012/01/jan-26-1906-fred-marriott-lets-off-some-steam/.

58 **"There is something uncanny":** "Topics of the Times," *New York Times*, January 3, 1899, 8.

59 **"cars," from the old Latin word:** *Medieval Latin: An Introduction*

*and Bibliographical Guide* (Washington, D.C.: Catholic University of America Press, 1996).

59   **The most popular cars were:** Bryan Appleyard, *The Car: The Rise and Fall of the Machine That Made the Modern World* (New York: Simon & Schuster, 2022).

59   **yearly salary of the average worker:** *Census of Manufactures: 1905 Earnings of Wage-Earners* (Washington, D.C.: U.S. Government Printing Office, 1908).

59   **Ford Quadricycle, a four-wheeled:** Henry Ford and Samuel Crowther, *My Life and Work* (Garden City, NY: Doubleday, Page, 1922).

60   **"the great multitude":** Robert H. Casey, *The Model T: A Centennial History* (Baltimore: Johns Hopkins University Press, 2016).

60   **"It will be large enough for the family":** Ford and Crowther, *My Life and Work*.

61   **Once it was running:** Thomas Alvin Boyd, *Charles F. Kettering: A Biography* (Washington, D.C.: Beard Books, 2002).

62   **Alfred P. Sloan, the savvy CEO:** Alfred Pritchard Sloan and Boyden Sparkes, *Adventures of a White-Collar Man* (New York: Doubleday, Doran, 1941).

62   **Sloan also wanted more streams of revenue:** David Farber, *Sloan Rules: Alfred P. Sloan and the Triumph of General Motors* (Chicago: University of Chicago Press, 2002).

63   **Midgley settled into life:** Thomas Midgley, unpublished typescript, Midgley family archives. Reviewed by the author.

63   **He took two thin strips:** Sharon Bertsch McGrayne, *Prometheans in the Lab: Chemistry and the Making of the Modern World* (New York: McGraw-Hill, 2001).

64   **Germans were using a flammable kitchen detergent:** Williams Haynes, *American Chemical Industry: The Merger Era, 1923–1929* (United States: Van Nostrand, 1945).

64   **Midgley bent over to inspect:** McGrayne, *Prometheans in the Lab*.

64   **Midgley decided to experiment on himself:** Charles Franklin Kettering, *Memoir of Thomas Midgley, Jr., 1889–1944* (Washington, D.C.: National Academy of Sciences, 1947).

65   **he submitted an article about the incident:** "Works and Laboratory Accidents: A Novel Method of Removing Metal from an Eye; Fire Hazards in Dyestuff Storage and Mixing; Explosions with Ammoniacal Silver Oxide Solutions; Reports from the Chemical Laboratory of the Massachusetts District Police," *Journal of Industrial & Engineering Chemistry* 11, no. 9 (September 1919): 982–95, https://doi.org/10.1021/ie50117a017.

65 **Kettering asked Midgley to turn:** Kettering, *Memoir of Thomas Midgley.*

66 **the combustion engine had a series:** Harry Egerton Wimperis, *The Internal Combustion Engine: Being a Text Book on Gas, Oil and Petrol Engines for the Use of Students and Engineers* (London: Constable, 1908).

68 **so he called it "engine knock":** Letter from Lodge Brothers & Co. Ltd., *The Motor Cycle*, November 12, 1914.

68 **A British engineer named Harry Ricardo:** William Rede Hawthorne, "Harry Ralph Ricardo, 26 January 1885–18 May 1974," *Biographical Memoirs of Fellows of the Royal Society* 22 (November 1976): 358–80.

69 **On the octane "scale":** "Fact #940: Diverging Trends of Engine Compression Ratio and Gasoline Octane Rating," Office of Energy Efficiency and Renewable Energy, August 29, 2016, https://www .energy.gov/eere/vehicles/fact-940-august-29-2016-diverging -trends-engine-compression-ratio-and-gasoline-octane.

## Chapter 6

73 **The day Midgley realized this:** Thomas Midgley, *From the Periodic Table to Production: The Biography of Thomas Midgley, Jr., the Inventor of Ethyl Gasoline and Freon Refrigerants* (Corona, CA: Stargazer Publishing, 2001).

74 **Ethyl alcohol, the same substance:** William Kovarik, "The Ethyl Controversy" (PhD diss., University of Maryland, College Park, 1993).

74 **It worked so well that:** H. B. Dixon, "Researches on Alcohol as an Engine Fuel," *Society of Automotive Engineers Journal* (December 1920): 521.

74 **"fuel of the future":** Kovarik, "Ethyl Controversy."

75 **"Alcohol has tremendous advantages":** William Kovarik, "Ethyl: The 1920s Conflict over Leaded Gasoline & Alternative Fuels," presentation, American Society for Environmental History Annual Conference, Providence, RI, March 26, 2003.

75 **7 billion barrels of oil:** David White, "The Petroleum Resources of the World," *Annals of the American Academy of Political and Social Science* 89 (1920): 111–34.

75 **240 billion barrels of oil:** *U.S. Field Production of Crude Oil* (Washington, D.C.: U.S. Energy Information Administration, 2023).

75 **simply be a "bridge fuel":** Thomas Midgley, "The Application of Chemistry to the Conservation of Motor Fuels," *Journal of*

*Industrial & Engineering Chemistry* 14, no. 9 (September 1922): 849–51, https://doi.org/10.1021/ie50153a050.

76 **"Energy expended by the sun":** "World Looks to Sun for Auto Power," *Kansas City Post*, November 9, 1922, 16.

76 **"You know, fellows, a bean is pretty smart":** T. A. Boyd, *Professional Amateur: The Biography of Charles Franklin Kettering* (New York: Dutton, 1957), 192.

77 **Midgley wanted to dye gasoline:** Thomas Midgley Jr., "How We Found Ethyl Gas," *Motor Magazine* (January 1925): 92–94.

78 **"I doubt if humanity":** William Kovarik, "Charles F. Kettering and the 1921 Discovery of Tetraethyl Lead," paper presented at Society of Automotive Engineers, Fuels and Lubricants Division, Baltimore, MD, 1994. Original source used by Kovarik: Kettering to Midgley, September 14, 1920, Midgley files, unprocessed, General Motors Institute.

78 **Then he turned to bromine:** Alfred B. Garrett, "Lead Tetraethyl: Thomas Midgley, Jr., T. A. Boyd, and C. A. Hochwalt," *Journal of Chemical Education* 39, no. 8 (1962): 414, https://doi.org/10.1021/ed039p414.

78 **"Where have you *been*?":** Midgley, "How We Found Ethyl Gas."

79 **"greeting me with gas masks on":** Thomas Midgley Jr., "From the Periodic Table to Production," *Journal of Industrial & Engineering Chemistry* 29, no. 2 (1937): 241–44. Republished by Stargazer Publishing, Corona, CA, 2001.

79 **"satanified garlic odor":** Midgley, "How We Found Ethyl Gas."

79 **"a period function":** Ethyl Corporation Public Relations Department, *History of Ethyl Corporation*, July 1, 1951, 18.

79 **"a fox hunt":** Charles Franklin Kettering, *Memoir of Thomas Midgley, Jr., 1889–1944* (Washington, D.C.: National Academy of Sciences, 1947), 363.

80 **It had been around since the 1850s:** Dietmar Seyferth, "The Rise and Fall of Tetraethyllead. 2," *Organometallics* 22, no. 25 (2003): 5154–78, https://doi.org/10.1021/om030621b.

80 **not long after a German chemist:** Jamie Lincoln Kitman, "The Secret History of Lead," *Nation* 270, no. 11 (March 20, 2000), https://www.thenation.com/article/archive/secret-history-lead/.

80 **On December 9, 1921, they fed:** David Farber, *Sloan Rules: Alfred P. Sloan and the Triumph of General Motors* (Chicago: University of Chicago Press, 2002).

81 **"Listen! Hurry! Come quick!":** Rosamond McPherson Young, *Boss Ket: A Life of Charles F. Kettering* (New York: Longmans, Green, 1961), 153.

81   **143 chemicals, or 15,000, or possibly 33,000:** Young, *Boss Ket*, 154;
     John C. Lane of Ethyl Corp., "Gasoline and Other Motor Fuels," in
     *Encyclopedia of Chemical Technology* (New York: John Wiley & Sons,
     1980), 656.

82   **GM reported selling 193,000 cars:** Richard S. Tedlow, "The
     Struggle for Dominance in the Automobile Market: The Early Years
     of Ford and General Motors," *Business and Economic History* (papers
     presented at the thirty-fourth annual meeting of the Business
     History Conference), vol. 17 (1988), 49–62.

82   **"When Kettering found":** Testimony of Ferris E. Hurd, General
     Motors attorney, *United States v. du Pont*, 1953, 7986.

82   **Midgley weakened the amount:** Ethyl Corporation Public
     Relations, *History of Ethyl Corporation*, 21.

83   **"We've got to have a name":** Young, *Boss Ket*, 155.

83   **General Motors' attorneys applied for a patent:** "Patent
     1,592,954: Fuel," U.S. Patent Office, July 20, 1926.

84   **the Associated Press published a bulletin:** "News from the Auto
     World," *Roanoke (VA) Times,* February 26, 1922, 21.

84   **General Motors' stock was up:** "New York Stocks," *Palladium-Item*
     (Richmond, IN), February 20, 1922, 12; "New York Stocks," *Tampa
     Tribune,* December 31, 1922, 50.

## Chapter 7

85   **"very finicky," in her words:** Letter from Alice Hamilton to Jessie
     Hamilton, March 6, 1919, Schlesinger Library, Harvard University.

86   **she found a room in the home:** Letter from Alice Hamilton to Jessie
     Hamilton, March 6, 1919, Schlesinger Library, Harvard University.

86   **"I cannot think of one thing":** Letter from Alice Hamilton to
     Clara Landsberg, February 26, 1925, Schlesinger Library, Harvard
     University.

87   **Hamilton had settled into a nice rhythm:** Barbara Sicherman
     and Alice Hamilton, *Alice Hamilton: A Life in Letters* (Cambridge,
     MA: Harvard University Press, 1984), 258.

87   **"You almost perish":** Sicherman and Hamilton, *Alice Hamilton,* 259.

87   **"I felt I could not refuse":** Sicherman and Hamilton, *Alice
     Hamilton,* 259.

87   **In March she made two visits:** Alice Hamilton pocket datebook,
     1922, Schlesinger Library, Harvard University.

88   **in the back stacks of the John Crerar Library:** Letter from Alice
     Hamilton to unknown recipient, April 13, 1922, Schlesinger Library,
     Harvard University.

88    **her country house in Hadlyme, Connecticut:** Letter from Alice Hamilton to Agnes Hamilton, 1920, Schlesinger Library, Harvard University.

88    **"such friends could rarely":** Sicherman and Hamilton, *Alice Hamilton*, 26.

89    **She recalled the embarrassment she felt:** Sicherman and Hamilton, *Alice Hamilton*, 19.

89    **"cruel, neglectful, lying":** Letter from Alice Hamilton to Agnes Hamilton, April 28, 1922, Schlesinger Library, Harvard University.

89    **"There seems to be no valid reason":** Letter from Alice Hamilton to Margaret Hamilton, March 9, 1923, Schlesinger Library, Harvard University.

90    **"one of us":** Sicherman and Hamilton, *Alice Hamilton,* 199.

90    **the House of Truth, a cooperative:** Brad Snyder, *The House of Truth: A Washington Political Salon and the Foundations of American Liberalism* (Oxford: Oxford University Press, 2017).

90    **"The talk at the table":** Alice Hamilton, *Exploring the Dangerous Trades* (Ann Arbor: University of Michigan Press, 1947), 196.

91    **Hammar was so impressed with Hamilton's ideas:** Hamilton, *Dangerous Trades*, 9.

91    **What they really needed was money:** Hamilton, *Dangerous Trades*, 9.

91    **"socially conscious" . . . "scientific standpoint":** Letter from Alice Hamilton to Agnes Hamilton, November 10, 1921, Schlesinger Library, Harvard University.

92    **lead mining in the United States tripled:** National Minerals Information Center, "Lead—Historical Statistics (Data Series 140), U.S. Geological Survey," July 22, 2022, https://www.usgs.gov/media/files/lead-historical-statistics-data-series-140.

92    **The telltale sign was a bluish line:** Lydia Denworth, *Toxic Truth: A Scientist, a Doctor, and the Battle over Lead* (Boston: Beacon Press, 2008).

93    **"genuinely concerned" about:** Hamilton, *Dangerous Trades*, 9.

93    **Hammar also had close ties:** Peter Reich, *The Hour of Lead: A Brief History of Lead Poisoning in the United States over the Past Century and of Efforts by the Lead Industry to Delay Regulation* (Washington, D.C.: Environmental Defense Fund, 1992).

93    **$52,500 for a three-year study:** Reich, *Hour of Lead*, 14.

93    **there was only one catch:** Christopher C. Sellers, *Hazards of the Job: From Industrial Disease to Environmental Health Science* (Chapel Hill: University of North Carolina Press, 2000), 160.

94    **"What makes her performance":** Oral history taken by S. Benison, July 19, 1957, Holmes Hall, Countway Medical Library, Archives GA4, Box 14, pp. 171–72. Drawn from Reich, *Hour of Lead*, 14.

94    **Lawrence Fairhall collected samples:** Lawrence T. Fairhall, "Lead Studies: I. The Estimation of Minute Amounts of Lead in Biological Material," *Journal of Industrial Hygiene* 4, no. 1 (May 1922): 9–20.

95    **Fairhall had made two major discoveries:** A. S. Minot, "Lead Studies: II. A Critical Note on the Electrolytic Determination of Lead in Biological Material," *Journal of Biological Chemistry* 55, no. 1 (January 1923): 1–8, https://doi.org/10.1016/S0021-9258(18)85691-5.

95    **If lead was ingested or inhaled:** Joseph C. Aub, Paul Reznikoff, and Dorothea E. Smith, "Lead Studies: III. The Effects of Lead on Red Blood Cells; Part 1. Changes in Hemolysis," *Journal of Experimental Medicine* 40, no. 2 (July 31, 1924): 151–72, https://doi.org/10.1084/jem.40.2.151.

96    **At 1 part per million:** J. A. Key, "Lead Studies: IV. Blood Changes in Acute Lead Poisoning in Rabbits with Especial Reference to Stippled Cells," *American Journal of Physiology* 70 (January 1924): 86–99.

97    **money came in from the National Research Council:** Aub, Reznikoff, and Smith, "Lead Studies: III. The Effects of Lead on Red Blood Cells," 189.

97    **she enjoyed the part of research:** Alice Hamilton letters, 1919–1926, Schlesinger Library, Harvard University.

98    **debate between Hamilton and the suffragist Doris Stevens:** Alice Hamilton and Doris Stevens, "The Blanket Amendment—A Debate," *Forum* 72 (August 1924): 421–26.

98    **"To my mind":** Sicherman and Hamilton, *Alice Hamilton*, 256.

99    **she remained working on her book:** Alice Hamilton, *Industrial Poisons in the United States* (New York: Macmillan, 1925).

99    **"let drop valuable leads":** Hamilton, *Industrial Poisons*, iv.

99    **"I have not even opened":** Letter from Alice Hamilton to Clara Landsberg, May 31, 1925, Schlesinger Library, Harvard University.

100   **"Lead is the chief harmful agent":** Hamilton, *Industrial Poisons*, 286.

## Chapter 8

102   ***"When used, a remarkable change":*** Barney Oldfield, "Barney Oldfield Says," *Evening Herald* (Fall River, MA), July 29, 1922, 9.

102   **Midgley believed that if Ethyl:** Thomas Midgley, "Factory

Correspondence, Miami Beach, Florida," typewritten letter from Thomas Midgley to Thomas Kettering, March 2, 1923.

103  **overcome a few obstacles:** William J. Kovarik, "The Ethyl Controversy" (PhD diss., University of Maryland, College Park, 1993), 85.

103  **Midgley eventually came upon bromine:** Thomas Midgley Jr., "How We Found Ethyl Gas," *Motor Magazine* (January 1925): 92–94.

103  **two men in 1916 who promoted a pill:** "Inventor Convicted of Gasoline Fraud," *New York Times*, November 2, 1922.

103  **Midgley wondered if the pill concept:** William Kovarik, "Charles F. Kettering and the 1921 Discovery of Tetraethyl Lead," paper presented at Society of Automotive Engineers, Fuels and Lubricants Division, Baltimore, MD, 1994. Kovarik cites letter from F. O. Clements to H. E. Talbott, February 4, 1919, unprocessed Midgley files, General Motors Institute.

104  **When Midgley sent out samples:** Kovarik, "Ethyl Controversy," 86.

104  **Midgley to receive the William H. Nichols medal:** "Nichols Medal Award," *Journal of Industrial & Engineering Chemistry* 15, no. 4 (April 1923): 421.

104  **Barney Oldfield's column:** "Better Fuel Often Cures Engine Knocking," *Arkansas Democrat* (Little Rock), July 30, 1922, 15.

104  **Columbia, South Carolina, *Sunday Record*:** "Better Fuel for the Motorist," *Sunday Record* (Columbia, SC), July 30, 1922, 3.

106  **"The demonstration got entirely":** Testimony of W. F. Harrington, *United States of America v. E. I. du Pont de Nemours and Company, General Motor Corporation, United States Rubber Company, Christiana Securities Company, Delaware Realty & Investment Corporation, Pierre S. du Pont, Lammot du Pont, Irenee du Pont*, March–July 1950, court case notes, National Archives, Chicago.

106  **Midgley was performing another demonstration:** Kovarik, "Ethyl Controversy," 86. Kovarik cites letter from Midgley to Dr. R. L. Allen, September 9, 1922, unprocessed Midgley files, General Motors Institute.

107  **"After about a year's work":** Kovarik, "Ethyl Controversy," 87. Kovarik cites letter from Midgley to H. N. Gilbert, January 19, 1923, unprocessed Midgley files, General Motors Institute.

107  **Lead was known to inhibit learning:** Sarah E. Royce, Herbert L. Needleman, U.S. Agency for Toxic Substances and Disease Registry, and DeLima Associates, *Lead Toxicity* (Atlanta: U.S. Department of Health & Human Services, Public Health Service, Agency for Toxic Substances and Disease Registry, 2000), 17.

108   **"It would not surprise me"**: Kovarik, "Ethyl Controversy," 87. Kovarik cites letter from Midgley to A. W. Browne, December 2, 1922, unprocessed Midgley files, General Motors Institute.

108   **letters from experts who knew**: T. A. Boyd, "The Early History of Ethyl Gasoline," Research Laboratory Division, General Motors Corporation, Detroit, June 8, 1943, 190.

108   **"creeping and malicious poison"**: Jamie Lincoln Kitman, "The Secret History of Lead," *Nation* 270, no. 11 (March 20, 2000), https://www.thenation.com/article/archive/secret-history-lead/.

109   **The foremost of the four experts was Yandell Henderson**: John B. West, "Yandell Henderson, April 23, 1873—February 18, 1944," *Biographical Memoirs*, vol. 74 (Washington, D.C.: National Academy Press, 1998), 144–58.

109   **Henderson was so alarmed**: D. Rosner and G. Markowitz, "A 'Gift of God'?: The Public Health Controversy over Leaded Gasoline During the 1920s," *American Journal of Public Health* 75, no. 4 (April 1985): 344–52, https://doi.org/10.2105/ajph.75.4.344.

109   **Cumming wrote back to Henderson**: General Hugh S. Cumming letters, 1921–1941, Center for the History of Medicine, Harvard University.

110   **in 1922, the presidency was held**: "Republican Party Platform of 1924," June 10, 1924, posted online by Gerhard Peters and John T. Woolley, American Presidency Project, n.d., https://www.presidency.ucsb.edu/node/273375.

110   **Harding believed that economic growth**: John A. Morello, *Selling the President, 1920: Albert D. Lasker, Advertising, and the Election of Warren G. Harding* (Westport, CT: Greenwood, 2001).

111   **"daily intake of minute quantities"**: Kitman, "Secret History of Lead."

111   **"the average street"**: Kitman, "Secret History of Lead."

112   **"My dear boss"**: Typewritten letter from Thomas Midgley to Charles Kettering, March 2, 1923, "Factory Correspondence, Miami Beach, Florida."

## Chapter 9

115   **motorists who stopped to fill their cars**: T. A. Boyd, "The Early History of Ethyl Gasoline," Research Laboratory Division, General Motors Corporation, Detroit, June 8, 1943, 195.

116   **a GM employee started to market**: Stuart W. Leslie, *Boss Kettering* (New York: Columbia University Press, 1983).

117   **"They did not realize"**: Testimony of Charles F. Kettering, *United States of America v. E. I. du Pont de Nemours and Company, General*

*Motor Corporation, United States Rubber Company, Christiana Securities Company, Delaware Realty & Investment Corporation, Pierre S. du Pont, Lammot du Pont, Irenee du Pont*, March–July 1950, court case notes, National Archives, Chicago, 3565.

117 **Indianapolis five-hundred-mile race:** Jamie Lincoln Kitman, "The Secret History of Lead," *Nation* 270, no. 11 (March 20, 2000), https://www.thenation.com/article/archive/secret-history-lead/.

117 **Midgley made sure reporters heard:** Leslie, *Boss Kettering*.

117 **"*Summer's comin', folks*":** "Diamond Gas Station," 1935, image 15907, Wisconsin Historical Society.

118 **fuel as a "wine color":** Boyd, "Early History of Ethyl Gasoline," 218.

118 **a new subsidiary within the company:** Boyd, "Early History of Ethyl Gasoline," 205.

118 **Alice Hamilton wrote to Surgeon General Cumming:** Letters from Hugh Cumming to Alice Hamilton, May 1923, Schlesinger Library, Harvard University.

119 **"cling to the pleasantness":** Barbara Sicherman and Alice Hamilton, *Alice Hamilton: A Life in Letters* (Cambridge, MA: Harvard University Press, 1984), 270.

119 **feverish pace of travel:** Alice Hamilton pocket datebooks, 1923, Schlesinger Library, Harvard University.

119 **mercury mines in Fresno and Salinas:** Alice Hamilton, *Exploring the Dangerous Trades* (Ann Arbor: University of Michigan Press, 1947).

120 **a rickshaw accident on a busy street:** "Jane Addams's Emergency Breast Surgery (1923)," Jane Addams Digital Edition, Jane Addams Papers Project, accessed June 9, 2024, https://digital.janeaddams.ramapo.edu/items/show/22115.

120 **"CAN COME IF NEEDED":** Madeleine Parker Grant, *Alice Hamilton: Pioneer Doctor in Industrial Medicine* (London: Abelard-Schuman, 1967), 182.

121 **took the RMS *Empress of Russia*:** "Australia Is Coming In with Seven Hundred Passengers from Orient," *Victoria (British Columbia) Daily Times,* July 11, 1923, 13.

122 **the "sunshine belt":** Pacific Mail Trans-Pacific Service advertisement, *Pacific Marine Review* (San Francisco), January 1923, 616.

122 **pile of letters from fellow doctors:** Letter from Alice Hamilton to Edith Hamilton, September 21, 1923, Schlesinger Library, Harvard University.

122 **a half-hearted study:** Boyd, "Early History of Ethyl Gasoline," 264.

**123** **he stuck his fingers into a beaker:** William J. Kovarik, "The Ethyl Controversy" (PhD diss., University of Maryland, College Park, 1993), 110.

**123** **Midgley reported these episodes:** Leslie, *Boss Kettering*, 165.

**123** **"for the purpose of increasing":** Joseph C. Robert, *Ethyl: A History of the Corporation and the People Who Made It* (Charlottesville: University Press of Virginia, 1983), 121.

**125** **"comment and criticism":** Boyd, "Early History of Ethyl Gasoline," 266.

**125** **yearly budget of barely $1 million:** *The Budget for the Service of the Fiscal Year Ending June 30, 1923* (Washington, D.C.: U.S. Government Printing Office, 1923).

**126** **The government investigation of tetraethyl leaded:** *Exhaust Gases from Engines Using Ethyl Gasoline* (Bruceton, PA: U.S. Bureau of Mines, 1924), 1–23.

**127** **the Ethyl Gas hounds:** Boyd, "Early History of Ethyl Gasoline," 267.

**128** **In phase two, the male subjects:** *Exhaust Gases from Engines Using Ethyl Gasoline*, 10.

**129** **"that there is no danger":** David Rosner and Gerald Markowitz, *Dying for Work: Workers' Safety and Health in Twentieth-Century America* (Bloomington: Indiana University Press, 1987), 124.

## Chapter 10

**131** **McSweeney started feeling queasy:** Mary Ross, "The Standard Oil's Death Factory," *Nation*, November 26, 1924, 561–62.

**132** **"looney gas building":** "Odd Gas Kills One, Makes Four Insane," *New York Times*, October 27, 1924, 1.

**132** **make tetraethyl lead faster using ethyl chloride:** N. P. Wescott, "Origins and Early History of the TetraEthyl Lead Business," DuPont corporate report, June 9, 1936, 1–6.

**132** **Ethyl chloride was a bigger molecule:** Joseph C. Robert, *Ethyl: A History of the Corporation and the People Who Made It* (Charlottesville: University Press of Virginia, 1986), 119.

**132** **fiftyfold increase in Ethyl sales:** Wescott, "Origins and Early History," 10.

**133** **When the drain backed up:** William J. Kovarik, "The Ethyl Controversy" (PhD diss., University of Maryland, College Park, 1993), 144.

**133** **eighty-five cents per hour:** Ross, "The Standard Oil's Death Factory," 562.

133 **twenty cents more than laborers:** "Odd Gas Kills One."

133 **McSweeney was shouting and acting:** "Odd Gas Kills One."

133 **another man working on the conjoined cone:** "Another Man Dies from Insanity Gas," *New York Times*, October 28, 1924, 1.

134 **all three men were dead:** "Third Victim Dies from Poison Gas," *New York Times*, October 29, 1924, 23.

134 **medical examiners in New Jersey and New York:** Christian Warren, *Brush with Death: A Social History of Lead Poisoning* (Baltimore: Johns Hopkins University Press, 2001), 119.

135 **"ODD GAS KILLS ONE":** "Odd Gas Kills One."

135 **"MYSTERY GAS CRAZES":** "Mystery Gas Crazes 12 in Laboratory," *New York Herald Tribune*, October 28, 1924, 1.

135 **"GAS MADNESS STALKS":** "Gas Madness Stalks Plant; 2 Die, 3 Crazed," *New York World*, October 28, 1924, 1.

136 **Some of the papers called the gas:** William J. Kovarik, "The Ethyl Conflict & the Media," paper presented at the Association for Education in Journalism and Mass Communication, April 1994.

136 **"From my observation":** "Odd Gas Kills One."

137 **"picking through the maze":** "Ethyl Gas Sale Stopped Today by Standard Oil," *New York Herald Tribune*, October 28, 1924, 1.

137 **"the dipping of one's finger":** Kovarik, "Ethyl Controversy," 110.

137 **Mann was exposed to the same:** "Chief Chemist Escapes as Loony Gas Victim," *Brooklyn Daily Eagle*, November 3, 1924, 1.

137 **"These men probably went insane":** "Odd Gas Kills One."

138 **"Otherwise [my son] would have quit":** "Odd Gas Kills One."

138 **They were in a "funk":** *United States v. du Pont*, U.S. District Court, Chicago, November 18, 1952, 126 F. Supp. 235, trial testimony, 2169.

139 **"life, health, and reason":** "Another Man Dies from Insanity Gas."

140 **"If an automobile":** "Another Man Dies from Insanity Gas."

140 **He bought five goats:** Rosamond McPherson Young, *Boss Ket: A Life of Charles F. Kettering* (New York: Longmans, Green, 1961), 163.

141 **reporters turned to Charles Norris:** C. Norris and A. O. Gettler, "Poisoning by Tetra-Ethyl Lead: Postmortem and Chemical Findings," *Journal of the American Medical Association* 85, no. 11 (1925): 818–20.

142 **"There was an astonishingly large":** "Gaseous Lead Poisoning," in *American Medicine*, vol. 30, ed. H. Edwin Lewis (Burlington, VT: American-Medicine Publishing, 1924), 626.

# Chapter 11

144 **The story spread quickly and evolved:** Rosamond McPherson Young, *Boss Ket: A Life of Charles F. Kettering* (New York: Longmans, Green, 1961), 162.

144 **"Something like five men":** William J. Kovarik, "The Ethyl Controversy" (PhD diss., University of Maryland, College Park, 1993), 147.

145 **"Doesn't that prove":** Young, *Boss Ket,* 163.

145 **"Under proper safeguards":** "Bar Ethyl Gasoline as 5th Victim Dies," *New York Times,* October 31, 1924, 1.

145 **Midgley asked an assistant:** David Rosner and Gerald Markowitz, *Dying for Work: Workers' Safety and Health in Twentieth-Century America* (Bloomington: Indiana University Press, 1987), 130.

146 **"I'm not taking any chance":** "Bar Death Gas in City as 5th Victim Dies," *New York Herald Tribune,* October 31, 1924, 1.

146 **wasn't it true that two other men died:** Kovarik, "Ethyl Controversy," 112.

146 **"This extremely dilute product":** "Use of Ethylated Gasoline Barred Pending Inquiry," *New York World,* October 31, 1924, 1.

147 **"Without desiring to attach":** Mary Ross, "The Standard Oil's Death Factory," *Nation,* November 26, 1924, 562.

147 **"the men, regardless of warnings":** Rosner and Markowitz, *Dying for Work,* 125.

147 **"a man's undertaking":** "Bar Death Gas in City as 5th Victim Dies."

147 **such as glycerin:** William J. Kovarik, "The Ethyl Conflict & the Media," paper presented at the Association for Education in Journalism and Mass Communication, April 1994.

148 **the attorney general assembled a grand jury:** "Absolved of 5 Gas Deaths," *New York Times,* February 12, 1925, 13.

148 **The Gulf Refining Company:** T. A. Boyd, "The Early History of Ethyl Gasoline," Research Laboratory Division, General Motors Corporation, Detroit, June 8, 1943, 272.

148 **"until it can be demonstrated":** "Announcement," *Pittsburgh Post,* November 1, 1923, 11.

149 **Lee pioneered a way:** Ray Eldon Hiebert, *Courtier to the Crowd: The Story of Ivy Lee and the Development of Public Relations* (Ames: Iowa State University Press, 1968), 37.

149 **a deadly coal mine strike:** Ron Chernow, *Titan: The Life of John D. Rockefeller, Sr.* (New York: Knopf Doubleday, 2004), 584.

150 **Lee suggested that Midgley should admit:** Hiebert, *Courtier to the Crowd*, 268.

152 **he made his way to the health boards:** Boyd, "Early History of Ethyl Gasoline," 272.

153 **Bureau of Mines sent the media its Ethyl-approved report:** R. R. Sayers et al., *Exhaust Gases from Engines Using Ethyl Gasoline* (Washington, D.C.: Department of Commerce, Bureau of Mines, 1924).

154 **"NO PERIL TO PUBLIC":** "No Peril to Public Seen in Ethyl Gas," *New York Times*, November 1, 1924, 17.

## Chapter 12

157 **"responsible industries":** Letter from Alice Hamilton to family from Moscow, October 19, 1924, Papers of the Hamilton family, 1818–1974, MC 278: M-24, Schlesinger Library, Harvard University.

158 **"The women do not waste time":** Letter from Alice Hamilton to family from Moscow, October 19, 1924.

158 **become a "media storm":** Letter from Clara Landsberg to Alice Hamilton, November 6, 1924, Schlesinger Library, Harvard University.

158 **She scribbled the symptoms:** Alice Hamilton pocket notebook, 1924, Schlesinger Library, Harvard University.

158 **"This is a very dangerous form of lead":** Alice Hamilton, *Exploring the Dangerous Trades* (Ann Arbor: University of Michigan Press, 1947).

160 **"cast[s] doubt on negative results":** Alice Hamilton to Hugh S. Cumming, February 12, 1925, PHS File 1340, U.S. National Archives, Washington, D.C. From William J. Kovarik, "The Ethyl Controversy" (PhD diss., University of Maryland, College Park, 1993), 194.

161 **a young researcher named Paul Reznikoff:** Finding Aid to the Paul Reznikoff, MD (1896–1984), papers, 1922–1984, Medical Center Archives of New York–Presbyterian/Weill Cornell, 2006.

161 **snow for three days:** "Colder Weather to Arrive Today," *Boston Globe*, December 9, 1924, 1.

162 **"As an investigation":** *Journal of Industrial Hygiene and Abstract of the Literature*, official organ of the American Association of Industrial Physicians and Surgeons (Cambridge, MA: Harvard Medical School, 1925), 95.

163  **Hamilton was "pleased and collegial":** Alice Hamilton pocket datebook, December 1924, Schlesinger Library, Harvard University.

163  **"The amount of lead was kept":** "Tetra-ethyl Lead," *Journal of the American Medical Association* 84, no. 20 (May 16, 1925): 1484, https://doi.org/10.1001/jama.1925.02660460017008.

163  **She felt angry:** Letter from Alice Hamilton to Edith Hamilton, December 14, 1924, Schlesinger Library, Harvard University.

164  **the company employed several hundred women:** Mark Neuzil and Bill Kovarik, *Mass Media & Environmental Conflict: America's Green Crusades* (Thousand Oaks, CA: Sage Publications, 1996), 32–52.

165  **By 1925, after three women had died:** Kate Moore, *The Radium Girls: The Dark Story of America's Shining Women* (Naperville, IL: Sourcebooks, 2017).

165  **New Jersey Consumers League asked Hamilton:** Hamilton, *Dangerous Trades*, 416.

165  **Drinker at Harvard conducted:** William B. Castle, Katherine R. Drinker, and Cecil K. Drinker, "Necrosis of the Jaw in Radium Workers," *Journal of Industrial Hygiene* (August 1925): 373.

166  **"every girl" . . . "Do you suppose":** Neuzil and Kovarik, *Mass Media & Environmental Conflict*, 38.

166  **Johns Manville company, which ferociously defended:** Jessica van Horssen, *A Town Called Asbestos: Environmental Contamination, Health, and Resilience in a Resource Community* (Vancouver: UBC Press, 2016).

167  **Hamilton sat down to write to Walter Lippmann:** Letter from Alice Hamilton to Walter Lippmann, December 16, 1924, Lippmann Collection, Yale University Library.

168  **Woodrow Wilson appointed him in 1920:** Mike Stobbe, *Surgeon General's Warning: How Politics Crippled the Nation's Doctor* (Oakland: University of California Press, 2014), 52–53.

168  **He was tall and thin:** Stobbe, *Surgeon General's Warning*, 53.

168  **"Few men are better known":** "Medicine," *Time*, December 17, 1928, 22.

169  **$100,000 in settlement payments:** Memorandum of meeting of board of directors of Ethyl Gasoline Corp. by Irénée du Pont, December 23, 1924, government trial exhibit no. 676, *United States v. du Pont*. From Kovarik, "Ethyl Controversy," 151.

169  **It was Christmas Eve 1924:** Kovarik, "Ethyl Controversy," 152.

170  **"the desirability of having":** David Rosner and Gerald

Markowitz, *Dying for Work: Workers' Safety and Health in Twentieth-Century America* (Bloomington: Indiana University Press, 1987), 127.

171 **It rained all day:** "Weather," *Evening Star,* Washington, DC., December 24, 1924, 1.

## Chapter 13

175 **temperatures in January dropped below zero:** "Mercury May Hit Zero Mark Before Dawn," *Marion (OH) Star*, January 27, 1925, 1.

175 **negative 40 degrees:** "Cold Wave Is Moving Across State in Path of Heavy Snow Storm," *Troy (OH) Daily News*, January 27, 1925, 1.

176 **whom they called "the boys":** Alfred P. Sloan to Irénée du Pont, March 28, 1925, government trial exhibit no. 678, *United States v. du Pont et al.*, U.S. District Court, Chicago, 1953. From William J. Kovarik, "The Ethyl Controversy" (PhD diss., University of Maryland, College Park, 1993), 156.

176 **"it was a great mistake":** Kovarik, "Ethyl Controversy," 156.

176 **"bright" and "increasingly assured":** N. P. Wescott, "Origins and Early History of the TetraEthyl Lead Business," DuPont corporate report, June 9, 1936, 26.

177 **the monthly profits had dropped:** Wescott, "Origins and Early History," 26.

178 **Kettering contacted Kehoe in August:** Christian Warren, *Brush with Death: A Social History of Lead Poisoning* (Baltimore: Johns Hopkins University Press, 2001), 129.

178 **"It has been shown experimentally":** Robert A. Kehoe, "Tetra-Ethyl Lead Poisoning: Clinical Analysis of a Series of Nonfatal Cases," *Journal of the American Medical Association* 85, no. 2 (July 11, 1925): 108–10, https://doi.org/10.1001/jama.1925.02670020028014.

179 **known as the Kehoe principle:** Jerome O. Nriagu, "Clair Patterson and Robert Kehoe's Paradigm of 'Show Me the Data' on Environmental Lead Poisoning," *Environmental Research* 78, no. 2 (August 1998): 71–78, https://doi.org/10.1006/enrs.1997.3808.

180 **the city's tradition of tabloid reporting:** John D. Stevens, *Sensationalism and the New York Press* (New York: Columbia University Press, 1991).

180 **The *New York World*, however, was a holdover:** George Juergens, *Joseph Pulitzer and the* New York World (Princeton, NJ: Princeton University Press, 2015).

181 **Pulitzer's *World* was advocating for:** John Langdon Heaton, *The Story of a Page: Thirty Years of Public Service and Public Discussion in the Editorial Columns of the* New York World (New York: Harper & Brothers, 1913).

181 **"crusade" against Ethyl:** "Dangerous Leaded Gasoline (Tetraethyl) Sale Stopped After Fight by *World*," *New York World*, May 5, 1925, 1.

181 **"Given the shortcomings":** Letter from Alice Hamilton to Walter Lippmann, February 11, 1925, Beinecke Library, Yale University.

182 **"Nobody has ever":** Walter Lippmann, *Men of Destiny* (New York: Macmillan, 1927), 13.

182 **In February 1925, it reported:** Kovarik, "Ethyl Controversy," 124.

182 **"The experiments conducted":** "Need for a Prompt Investigation," *New York World*, April 12, 1925, 2E.

183 **"fuel of the future":** "Recent Patents on Mixed Fuels," *Scientific American* 124, no. 24 (December 11, 1920): 593.

184 **"so very interesting":** Telegram from Charles F. Kettering to Alfred P. Sloan, 1924, General Motors Institute. From Kovarik, "Ethyl Controversy," 149.

184 **"injustice . . . wasn't given all of the facts":** Alice Hamilton's datebook, 1924, Schlesinger Library, Harvard University.

185 **"industrial hygiene," was woefully:** American Industrial Hygiene Association advertisement, "The Future of AIHA Has Arrived," in *Safety+Health* 201, no. 6 (June 30, 2020): 14.

186 **Midgley wrote to Kettering:** Thomas Midgley, unpublished typescript, Midgley family archives. Reviewed by the author.

## Chapter 14

190 **On Cumming's desk were letters:** Correspondence from 1924 to 1925. Records of the Public Health Service, 1912–1968, 90.2.1, National Archives, College Park, MD.

190 **Hamilton visited Cumming twice:** Alice Hamilton's pocket datebook, 1925, Schlesinger Library, Harvard University.

190 **"I observe it":** Letter from Alice Hamilton to Hugh Cumming, March 6, 1925, Records of the Public Health Service, 1912–1968, 90.2.1, National Archives, College Park, MD.

190 **"It is the question whether scientific":** "Sees Deadly Gas a Peril in Streets," *New York Times*, April 22, 1925, 26.

191 **Cumming had a broad philosophy:** Mike Stobbe, *Surgeon General's Warning: How Politics Crippled the Nation's Doctor* (Oakland: University of California Press, 2014), 56.

191 **warn of the dangers of smoking:** Stobbe, *Surgeon General's Warning*, 58.

191 **infected oysters in the Chesapeake Bay:** "Lifting Oyster Ban Within Few Days Probable," *Baltimore Sun*, January 31, 1925, 20.

191 **millions of rats destroying the state's farms:** "Health Inspector Here," *Los Angeles Daily News*, January 29, 1925, 3.

192 **"investigate such questions":** U.S. Public Health Service, *Proceedings of a Conference to Determine Whether or Not There Is a Public Health Question in the Manufacture, Distribution, or Use of Tetraethyl Lead Gasoline* (Washington, D.C.: U.S. Government Printing Office, 1925), 3.

192 **Mellon sensed it could be awkward:** C. G. Sarkar, "Tetraethyllead (TEL) in Gasoline as a Case of Contentious Science and Delayed Regulation: A Short Review," *Oriental Journal of Chemistry* 36, no. 1 (2020), http://dx.doi.org/10.13005/ojc/360111.

192 **"had not been asked":** "Synthetic Marvels Arouse Scientists," *New York Times*, May 8, 1925, 21.

192 **report on alternative anti-knock fuels:** William J. Kovarik, "The Ethyl Controversy" (PhD diss., University of Maryland, College Park, 1993), 132.

193 **to demote Midgley and Kettering:** Kovarik, "Ethyl Controversy," 156.

193 **as accidents continued to happen:** Testimony of Charles F. Kettering, *United States v. du Pont*, 3565.

194 **du Pont had signed a licensing deal:** Kovarik, "Ethyl Controversy," 306.

194 **"violently opposed" to being fired:** Alfred P. Sloan to Pierre du Pont, March 28, 1925, government trial exhibit no. 678, *United States v. du Pont et al.*, U.S. District Court, Chicago, 1953. From Kovarik, "Ethyl Controversy," 156.

194 **Kettering and Midgley asked that their demotions:** Joseph C. Robert, *Ethyl: A History of the Corporation and the People Who Made It* (Charlottesville: University Press of Virginia, 1986), 124–25.

195 **Ethyl ought to suspend all production:** "Dangerous Leaded Gasoline (Tetraethyl) Sale Stopped After Fight by *World*," *New York World*, May 5, 1925, 1.

195 **Midgley from continuing to defend:** Thomas Midgley, unpublished typescript, Midgley family archives. Reviewed by the author.

195 **"So far as science knows":** "Radium Derivative $5,000,000 an Ounce," *New York Times*, April 7, 1925, 23.

195 **making it a "bald-faced lie":** Kovarik, "Ethyl Controversy," 206.

197 **"Youngstown is hopelessly ugly":** Barbara Sicherman and Alice Hamilton, *Alice Hamilton: A Life in Letters* (Cambridge, MA: Harvard University Press, 1984), 288.

197 **film called *The Thundering Herd*:** "Thundering Herd: A Vital Western," *Exhibitors Herald*, February 28, 1925, 74.

198 **brief stays in Chicago and New York:** Alice Hamilton pocket datebook, 1925, Schlesinger Library, Harvard University.

198 **"Because of the enormous":** A. Hamilton, P. Reznikoff, and G. M. Burnham, "Tetra-Ethyl Lead," *Journal of the American Medical Association* 84, no. 20 (1925): 1486.

198 **more than 2,000 workers:** *Record of Industrial Accidents in the United States to 1925*, Bulletin of the United States Bureau of Labor Statistics, no. 425 (Washington, D.C.: U.S. Government Printing Office, January 1927), 30–35.

199 **"Publicity is a wonderful thing":** Letter from Alice Hamilton to Katherine Bowditch Codman, May 17, 1925, Schlesinger Library, Harvard University.

## Chapter 15

201 **every expert on every side:** U.S. Public Health Service, *Proceedings of a Conference to Determine Whether or Not There Is a Public Health Question in the Manufacture, Distribution, or Use of Tetraethyl Lead Gasoline* (Washington, D.C.: U.S. Government Printing Office, 1925).

201 **she always liked to be early:** Letter from Alice Hamilton to Agnes Hamilton, February 23, 1916, Schlesinger Library, Harvard University.

202 **the thirty-seven-room house:** *Annual Report of the Superintendent, United States Coast and Geodetic Survey, to the Secretary of Commerce for the Fiscal Year Ended June 30, 1919* (Washington, D.C.: U.S. Government Printing Office, 1919), 17–19.

202 **parquet floors, hardwood moldings, and wainscot paneling:** "Appraisement of the Butler Building," Congressional Serial Set, 50th Congress, 1st Session, House of Representatives, Ex. Doc. 262 (Washington, D.C.: U.S. Government Printing Office, 1889), 2.

203 **placing the chairs in rows:** U.S. Public Health Service, *Proceedings of a Conference,* 3.

203 **"I will ask the meeting to come to order!":** And all other quotes during USPHS conference, U.S. Public Health Service, *Proceedings of a Conference,* 3.

208 **"the minimizing of what seemed":** Letter from Alice Hamilton to Katherine Bowditch Codman, June 2, 1925, Schlesinger Library, Harvard University.

208 **"It has been clearly shown":** "Shift Ethyl Inquiry to Surgeon General," *New York Times*, May 21, 1925, 7.

213 **"are nothing but a murderer"**: T. A. Boyd, "The Early History of Ethyl Gasoline," Research Laboratory Division, General Motors Corporation, Detroit, June 8, 1943, 276.

219 **she caught a train to Schenectady:** Alice Hamilton pocket datebook, 1925, Schlesinger Library, Harvard University.

## Chapter 16

221 **"great progress" . . . "blaze of publicity":** Alice Hamilton, "What Price Safety: Tetra-ethyl Lead Reveals a Flaw in Our Defenses," *Survey Midmonthly* 54, no. 6 (June 15, 1925); republished in *Journal of Occupational Medicine* 14, no. 2 (February 1972): 98–100.

222 **The investigation finally started:** H. S. Cumming, "Report of Surgeon General's Committee on Tetraethyl Lead," *Journal of Industrial & Engineering Chemistry* 18, no. 2 (1926): 193–96.

224 **"half-baked report":** William J. Kovarik, "The Ethyl Controversy" (PhD diss., University of Maryland, College Park, 1993), 176.

224 **Winslow feared the issue:** David Rosner and Gerald Markowitz, "A 'Gift of God'?: The Public Health Controversy over Leaded Gasoline During the 1920s," *American Journal of Public Health* 75, no. 4 (April 1985): 344–52.

224 **"a more extensive study":** Rosner and Markowitz, "A 'Gift of God'?," 344–52.

225 **Leslie blamed the breakdown:** Stuart W. Leslie, *Boss Kettering* (New York: Columbia University Press, 1983), 167.

225 **"no good grounds for prohibiting":** "Report No Danger in Ethyl Gasoline," *New York Times*, January 20, 1926, 13.

225 **had been resolved:** "Poison Gasoline Declared Safe for Sale Again," *New York World*, January 21, 1926, 33.

226 **"There is a firm difference":** Letter from Alice Hamilton to Howard [Haggard], undated, Schlesinger Library, Harvard University.

226 **"I doubt very much":** "Report No Danger in Ethyl Gasoline."

226 **"clean bill of health":** T. A. Boyd, "The Early History of Ethyl Gasoline," Research Laboratory Division, General Motors Corporation, Detroit, June 8, 1943, 250.

226 **had cost Ethyl $3 million:** Boyd, "Early History of Ethyl Gasoline," 285.

227 **signs at Standard, Amoco, and Esso:** Jamie Lincoln Kitman, "The Secret History of Lead," *Nation* 270, no. 11 (2000).

227 **"ETHYL IS BACK":** Boyd, "Early History of Ethyl Gasoline," 285.

**227** **called Ethyl a "super fuel":** "Associated Ethyl Gasoline" advertisement, *Napa Register*, December 16, 1926, 6.

**227** **with "no substitute":** "Knock Out That 'Knock'" advertisement, *Fort Worth Star-Telegram*, November 26, 1926, 15.

**227** **"faster pick-up":** "Associated Ethyl Gasoline" advertisement, *Fresno Bee*, December 30, 1926, 5.

**227** **society remaking itself around the automobile:** Rosner and Markowitz, "A 'Gift of God'?," 344–52.

**228** **Ethyl could only be manufactured:** Boyd, "Early History of Ethyl Gasoline," 282.

**228** **10 billion gallons of gasoline:** "Business Day by Day: U.S. Gasoline Consumption," *Pittsburgh Press*, March 12, 1927, 8.

**228** **rise further to 12 billion gallons:** "Gasoline Use Rises 10.4%," *New York Times*, March 11, 1937, 42.

**228** **more than 130 billion gallons:** U.S. Energy Information Administration, "Petroleum Monthly Supply Archives," June 30, 2023, https://www.eia.gov/petroleum/supply/monthly /archive/2023/2023_06/psm_2023_06.php.

**228** **Ethyl amounted to more than $300 million:** Kovarik, "Ethyl Controversy," 91.

**228** **a Navy airship fueled by Ethyl:** Peter Andrews, "Lighter Than Air," *American Heritage Invention & Technology* 9, no. 1 (Summer 1993): 9.

**229** **"free-selling" policy:** N. P. Wescott, "Origins and Early History of the TetraEthyl Lead Business," DuPont corporate report, June 9, 1936, 30.

**231** **a U.S. government report signed by Cumming:** Cumming letters to public health authorities in Argentina, Brazil, and Canada, July to November 1934, Public Health Service, RG90 Box 98, National Archives, Washington, D.C.

**231** **"infinitesimal doses of lead":** Norman Porritt, "Cumulative Effects of Infinitesimal Doses of Lead," *British Medical Journal* 2, no. 3680 (July 18, 1931): 92–94.

**231** **"no good grounds":** Kitman, "Secret History of Lead."

**232** **fifty cents . . . to just eighteen cents:** Boyd, "Early History of Ethyl Gasoline," 286–87.

## Chapter 17

**235** **Midgley bought a fifty-acre plot:** Sharon Bertsch McGrayne, *Prometheans in the Lab: Chemistry and the Making of the Modern World* (New York: McGraw-Hill, 2001), 95.

236  **his erratic behavior:** McGrayne, *Prometheans in the Lab,* 99.

237  **was doing his "hat trick":** McGrayne, *Prometheans in the Lab,* 102.

237  **Midgley on another quest:** Alfred B. Garrett, "Freon: Thomas Midgley and Albert L. Henne," *Journal of Chemical Education* 39, no. 7 (1962): 361.

237  **Making consistently cool air:** R. J. Thompson, "Freon, a Refrigerant," *Journal of Industrial & Engineering Chemistry* 24, no. 6 (1932): 620–23.

238  **"death gas ice boxes":** Mark D. Hersey and Theodore Steinberg, eds., *A Field on Fire: The Future of Environmental History* (Tuscaloosa: University of Alabama Press, 2019), 61.

238  **He needed a combination of elements:** McGrayne, *Prometheans in the Lab,* 97.

239  **"everything looked right":** McGrayne, *Prometheans in the Lab,* 97.

239  **he wowed audiences by inhaling:** "Earth Grapples with 'Killer' Chemicals, Once Called a Miracle," *Greensboro (NC) News & Record,* April 17, 1988, 61.

239  **railcars, movie theaters, restaurants:** McGrayne, *Prometheans in the Lab,* 97.

239  **General Motors marketed that "air-conditioning":** "Mental and Physical Well-Being Is Aided by Air Conditioning," *Hartford Courant,* June 30, 1935, 30.

240  **extend the average American lifespan:** "World Population Prospects 2022: United States," United Nations, Department of Economic and Social Affairs, Population Division, https://population.un.org/wpp/Graphs/Probabilistic/EX/BothSexes/840.

240  **Refrigeration allowed soda fountains:** Richard Osborn Cummings, *The American Ice Harvests: A Historical Study in Technology, 1800–1918* (Berkeley: University of California Press, 1949).

241  **Two scientists alerted the world:** Mario J. Molina and F. S. Rowland, "Stratospheric Sink for Chlorofluoromethanes: Chlorine Atom-Catalysed Destruction of Ozone," *Nature* 249 (1974): 810–12.

241  **"not based on authoritative evidence":** William Glaberson, "Behind Du Pont's Shift on Loss of Ozone Layer," *New York Times,* March 26, 1988, 41.

242  **cases of skin cancer and cataracts:** Cynthia Pollock Shea, "Why Du Pont Gave Up $600 Million," *New York Times,* April 10, 1988, section 3, p. 2.

242  **a unanimous treaty requiring the phaseout:** Donald Kaniaru, ed., *The Montreal Protocol: Celebrating 20 Years of Environmental Progress; Ozone Layer and Climate Protection* (London: Cameron May, 2007).

**242 back to its normal state by 2066:** *Scientific Assessment of Ozone Depletion: 2022*, United Nations Environment Programme, January 9, 2023, 1, https://www.unep.org/resources/publication/scientific -assessment-ozone-layer-depletion-2022.

**242 a different class of chemicals:** Mark O. McLinden and Marcia L. Huber, "(R)Evolution of Refrigerants," *Journal of Chemical & Engineering Data* 65, no. 9 (2020): 4176–93.

**242 the Environmental Protection Agency started requiring reductions:** Environmental Protection Agency final rule, *Phasedown of Hydrofluorocarbons: Establishing the Allowance Allocation and Trading Program Under the American Innovation and Manufacturing Act,* Document 86 FR 55116, *Federal Register* 86, no. 190 (October 5, 2020), https://www .federalregister.gov/documents/2021/10/05/2021-21030/ phasedown-of-hydrofluorocarbons-establishing-the -allowance-allocation-and-trading-program -under-the.

**243 "total phaseout" of CFC production:** *Stratospheric Ozone Depletion: Hearing Before the Subcommittee on Health and the Environment of the Committee on Energy and Commerce, House of Representatives, 101st Congress, 2nd Session, on H.R. 2699 . . . January 25, 1990* (Washington, D.C.: U.S. Government Printing Office, 1990), 334.

**243 "I'm doing something that's important":** Glaberson, "Behind Du Pont's Shift on Loss of Ozone Layer," 41, 43.

**243 "It shows that corporate America":** Shea, "Why Du Pont Gave Up $600 Million."

**243 called the company "very responsible":** Associated Press, "Du Pont Will Stop Making Ozone Killer: But Offers No Timetable for Ending Production," *Los Angeles Times*, March 24, 1988.

**243 building his mansion in Worthington:** McGrayne, *Prometheans in the Lab,* 91–104.

**244 fifty men to constantly improve:** Bill Arter, "Man-Made Caverns," *Columbia Dispatch Magazine*, October 31, 1965.

**245 "This comes out to be substantially equal":** Letter from Thomas Midgley to A. Ray Olpin, October 31, 1940, included in conference report of University of Utah Research Conference on the Identification of Creative Scientific Talent, Brighton, UT, August 27–30, 1955, 2.

**246 published in the *Journal of Industrial & Engineering Chemistry*:** T. Boyd, "Obituary. Thomas Midgley, Jr.," *Journal of the American Chemical Society* 75, no. 12 (1953): 2791–95.

**246 10,000 Americans caught polio:** C. C. Dauer, "Prevalence of

Poliomyelitis in the United States in 1940," *Public Health Reports* 56, no. 17 (1941): 875–83, https://doi.org/10.2307/4583711.

246 **a series of crossbeams and ropes:** McGrayne, *Prometheans in the Lab*, 104–5.

247 **adding more tetraethyl lead to American and British warplanes:** C. Boyden Gray and Andrew R. Varcoe, "Octane, Clean Air, and Renewable Fuels: A Modest Step Toward Energy Independence," *Texas Review of Law and Politics* 10 (2006): 18–19.

247 **he delivered his final speech:** Thomas Midgley, "Presidential Address," *Chemical & Engineering News Archive* 22, no. 19 (1944): 1646–49.

247 **Midgley killed himself:** Thomas Midgley death certificate, November 2, 1944, Columbus, Ohio, Department of Health.

247 **considered Midgley "like a son or a brother":** Charles Franklin Kettering, "Biographical Memoir of Thomas Midgley, Jr., 1889–1944," in *Biographical Memoirs*, vol. 24 (Washington, D.C.: National Academy of Sciences, 1947).

248 **Midgley's flourishes slowly faded:** Arter, "Man-Made Caverns."

248 **a circular highway, called the outerbelt:** File photo, "The "Outerbelt Expressway" (I-270) at Worthington, 1965," *Columbus Dispatch*, 2012.

## Chapter 18

250 **She awoke each morning at seven:** Letter from Alice Hamilton to Agnes Hamilton, April 27, 1932, Schlesinger Library, Harvard University.

250 **President's Research Committee on Social Trends:** Barbara Sicherman and Alice Hamilton, *Alice Hamilton: A Life in Letters* (Cambridge, MA: Harvard University Press, 1984), 321.

251 **"We have to approach":** Alice Hamilton, *Exploring the Dangerous Trades* (Ann Arbor: University of Michigan Press, 1947), 310.

251 **Infant mortality rates were sky-high:** Hamilton, *Dangerous Trades*, 311–12.

252 **"When it was first produced":** Alice Hamilton, *Industrial Toxicology* (New York: Harper & Brothers, 1934), 610.

252 **she labeled carbon tetrachloride:** Sicherman and Hamilton, *Alice Hamilton*, 325–31.

253 **"weakest of adversaries" . . . "I find it hard":** Letter from Alice Hamilton to lawyer Benjamin Cohen, November 5, 1932, Schlesinger Library, Harvard University.

254 **"I am getting old and garrulous":** Letter from Alice Hamilton to

Clara Landsberg, February 27, 1933, Schlesinger Library, Harvard University.

254 **she sent her article to be published:** Alice Hamilton, "Formation of Phosgene in Thermal Decomposition of Carbon Tetrachloride," *Journal of Industrial & Engineering Chemistry* 25, no. 5 (1933): 539–41.

254 **"misleading and inaccurate":** J. C. Olsen, "Discussion of 'Formation of Phosgene in Thermal Decomposition of Carbon Tetrachloride,' " *Journal of Industrial & Engineering Chemistry* 25, no. 5 (1933): 541–42.

254 **respiratory disease, kidney disease, liver damage:** *Hazards of Carbon Tetrachloride Fire Extinguishers: Recommended Practices Number 3* (Washington, D.C.: Federal Fire Council, 1967), 7–8.

254 **banned by the EPA in 1970:** *Toxicological Review of Carbon Tetrachloride*, EPA/635/R-08/005A, CAS No. 56-23-5 (Washington, D.C.: U.S. Environmental Protection Agency, May 2008).

255 **"I have never lost my confidence":** Charles F. Kettering, "Relation of Chemistry to the Individual," *Journal of Industrial & Engineering Chemistry* 25, no. 5 (May 1933): 484–86.

255 **two of the most memorable trips:** Hamilton, *Dangerous Trades*, chapter 20.

256 **"It is staggering to me":** Hamilton, *Dangerous Trades*, 382.

256 **"It was to me a dreadful thing":** Hamilton, *Dangerous Trades*, 400.

257 **a factory in a small Dutch village:** Hamilton, *Dangerous Trades*, 403.

257 **a small Royal typewriter:** Letter from Alice Hamilton to Jessie Hamilton, date unknown but 1943, Schlesinger Library, Harvard University.

258 **The FBI started following Hamilton:** Federal Bureau of Investigation file on Alice Hamilton, February–April 1954. Acquired by the author under Freedom of Information Act.

259 **"on constant watch for old poisons":** Madeleine Parker Grant, *Alice Hamilton: Pioneer Doctor in Industrial Medicine* (London: Abelard-Schuman, 1967), 206.

259 **named Woman of the Year in Medicine:** "Medicine: Woman of the Year," *Time*, November 19, 1956.

259 **Carson studied the subject:** Ralph H. Lutts, "Chemical Fallout: Rachel Carson's *Silent Spring*, Radioactive Fallout, and the Environmental Movement," *Environmental Review* 9, no. 3 (1985): 211–25, https://doi.org/10.2307/3984231.

260 **They called her a "hysterical woman":** " 'Silent Spring' Makes Protest Too Hysterical," *Arizona Daily Star* (Tucson), October 14, 1962, 37.

**260  they threatened lawsuits:** "Special Press Analysis of Rachel
Carson's 'Silent Spring,'" November 1 and December 6, 1962,
confidential reports and public relations department internal
memo, E. I. DuPont de Nemours, Hagley Museum and Library,
Wilmington, DE.

**260  until she turned ninety:** Sicherman and Hamilton, *Alice Hamilton*,
415.

**261  "an amazing lady":** "Dr. Alice Hamilton, Health Pioneer, Dies,"
*Hartford Courant*, September 23, 1970, 8.

**261  a "famous doctor":** "Rites Held for Famous Doctor," *Sidney (OH)
Daily News*, September 24, 1970, 1.

**261  "pioneer in medicine":** "Deaths Around the World," *Ventura
County (CA) Star*, September 24, 1970, 4.

**261  "a sterling example":** "Dr. Alice," *New London (CT) Day*, September
24, 1970, 20.

**261  Nixon heard of her death:** White House telegram, September 27,
1970, Schlesinger Library, Harvard University.

**261  "lasting contributions to the well-being":** "Dr. Alice Hamilton
Celebrates 100th Birthday," *New York Times*, February 28, 1969, 35.

**262  Hamilton its "founding mother":** "Celebrating the Life of Alice
Hamilton, Founding Mother of Occupational Medicine," *PBS
Newshour*, September 22, 2015, https://www.pbs.org/newshour
/health/celebrating-life-alice-hamilton-founding-mother
-occupational-medicine.

## Chapter 19

**264  collect pieces of lead ore:** Clair Patterson, "Interview with Clair
C. Patterson," interviewed by Shirley K. Cohen, March 5, 6, and 9,
1995 (1997), Caltech Oral Histories, Archives, California Institute of
Technology, Pasadena, https://oralhistories.library.caltech.edu/32/.

**264  He tried smaller samples:** Lydia Denworth, *Toxic Truth: A Scientist,
a Doctor, and the Battle over Lead* (Boston: Beacon Press, 2008).

**265  "I tracked back":** Denworth, *Toxic Truth*, 13.

**265  the cleanest two hundred square feet:** Douglas Smith, "Getting
the Lead Out," Human Resources, Caltech, September 21, 2015,
https://hr.caltech.edu/news/getting-lead-out-47935.

**265  he made his cleaning protocol:** Denworth, *Toxic Truth*, 19.

**266  having a heart attack:** Patrick N. Wyse Jackson,
"Geochronological Hits and Misses: Various Attempts to
Determine the Age of the Earth," *Open University Geological Society
Journal* 29, no. 2 (2008): 95.

**266** **collect water samples at different depths:** Denworth, *Toxic Truth,* 49.

**267** **to commit $30,000 per year to continue:** Patterson, "Interview with Clair C. Patterson."

**267** **"could readily be accounted for":** M. Tatsumoto and C. C. Patterson, "Concentrations of Common Lead in Some Atlantic and Mediterranean Waters and in Snow," *Nature* 199 (1963): 350–52.

**268** **"They tried to buy":** Denworth, *Toxic Truth,* 49.

**268** **American Petroleum Institute revoked its grant:** Patterson, "Interview with Clair C. Patterson."

**269** **80 micrograms per 100 grams of blood:** Denworth, *Toxic Truth,* 59.

**270** **"It's an example of how wrong":** Denworth, *Toxic Truth,* 66.

**271** **"Contaminated and Natural Lead Environments":** Clair C. Patterson, "Contaminated and Natural Lead Environments of Man," *Archives of Environmental Health: An International Journal* 11, no. 3 (1965): 344–60.

**271** **excerpted in the *New York Times*:** "Warning Is Issued on Lead Poisoning," *New York Times,* September 12, 1965, 71.

**271** **"noisy, angry, and sometimes incoherent":** Harriet Louise Hardy, *Challenging Man-Made Disease: The Memoirs of Harriet L. Hardy, M.D.,* ed. Emily W. Rabe (New York: Praeger, 1983), 121–22.

**271** **a reporter caught Kehoe by surprise:** Peter Reich, *The Hour of Lead: A Brief History of Lead Poisoning in the United States over the Past Century and of Efforts by the Lead Industry to Delay Regulation* (Washington, D.C.: Environmental Defense Fund, 1992), 38.

**271** **"What is your salary":** Hardy, *Challenging Man-Made Disease,* 122.

**272** **By 1966, the building was renamed:** "Our History: The History of the Department of Environmental and Public Health Sciences," University of Cincinnati College of Medicine, accessed September 11, 2023, https://med.uc.edu/depart/eh/about-us/history.

**272** **"childish ways of self assertion":** Letter from Robert Kehoe to Kenneth L. Kuykendall, June 7, 1973, Toxic Docs, Columbia University, Mailman School of Public Health, https://www.toxicdocs.org/d/byj4p4n4GQvEJB2zOGb6D3pm3?lightbox=1.

**273** **Muskie held hearings on the lead issue:** Herbert Needleman, "The Removal of Lead from Gasoline: Historical and Personal Reflections," *Environmental Research* 84, no. 1 (2000): 20–35, https://doi.org/10.1006/enrs.2000.4069.

**273** **twenty-one scientific studies about lead poisoning:** Reich, *The Hour of Lead,* 38.

273  **researchers found the real number:** Reich, *The Hour of Lead*, 40.

273  **the threshold was lowered to 25:** Denworth, *Toxic Truth*, xii.

274  **Blanchard used many of the same arguments:** "Requirement for Lead-Free Fuel May Be Delayed," *Los Angeles Times*, May 5, 1972, 3.

275  **regulate Ethyl a "witch hunt":** Lawrence E. Blanchard, "Washington Press Briefing," National Press Club, January 18, 1971.

275  **the government officials tasked with:** Lucas Reilly, "The Most Important Scientist You've Never Heard Of," *Mental Floss,* May 17, 2017, https://www.mentalfloss.com/article/94569/clair-patterson -scientist-who-determined-age-earth-and-then-saved-it.

275  **a new invention for cars:** *Consultant Report to the Committee on Motor Vehicle Emissions, Commission on Sociotechnical Systems, National Research Council, on an Evaluation of Catalytic Converters for Control of Automobile Exhaust Pollutants* (Washington, D.C.: U.S. Environmental Protection Agency, Office of Air and Waste Management, Office of Mobile Source Air Pollution Control, 1974).

276  **ten years for the full "phasedown":** Elin Hofverberg, "The History of the Elimination of Leaded Gasoline," Library of Congress Blogs, April 14, 2022, https://blogs.loc.gov/law/2022/04/the-history -of-the-elimination-of-leaded-gasoline/.

## Chapter 20

278  **Something struck him as odd:** David Rosner and Gerald Markowitz, "Standing Up to the Lead Industry: An Interview with Herbert Needleman," *Public Health Reports* 120, no. 3 (Spring 2005): 330–37.

278  **"They were gray and mute":** Lydia Denworth, *Toxic Truth: A Scientist, a Doctor, and the Battle over Lead* (Boston: Beacon Press, 2008), 36.

279  **Many of the kids had been exposed:** Rosner and Markowitz, "Standing Up to the Lead Industry," 332.

279  **Needleman published his findings in 1979:** Herbert L. Needleman et al., "Deficits in Psychologic and Classroom Performance of Children with Elevated Dentine Lead Levels," *New England Journal of Medicine* 300, no. 13 (1979): 689–95.

279  **"When somebody calls me":** Denworth, *Toxic Truth*, 136.

280  **A scientist in Baltimore tested levels:** Howard W. Mielke, "Lead in the Inner Cities: Policies to Reduce Children's Exposure to Lead May Be Overlooking a Major Source of Lead in the Environment," *American Scientist* 87, no. 1 (1999): 62–73.

280  **In New Zealand, someone tested:** Nick Wilson and John Horrocks, "Lessons from the Removal of Lead from Gasoline for

Controlling Other Environmental Pollutants: A Case Study from New Zealand," *Environmental Health* 7, article no. 1 (2008), https://doi.org/10.1186/1476-069x-7-1.

280 **when children are exposed to lead dust:** Carrie Arnold, "The Man Who Warned the World About Lead," *Nova*, May 31, 2017, https://www.pbs.org/wgbh/nova/article/herbert-needleman/.

281 **Rick Nevin noticed a strange coincidence:** Rick Nevin, "How Lead Exposure Relates to Temporal Changes in IQ, Violent Crime, and Unwed Pregnancy," *Environmental Research* 83, no. 1 (2000): 1–22, https://doi.org/10.1006/enrs.1999.4045.

281 **The legalization of abortion:** Kevin Drum, "Lead: America's Real Criminal Element," *Mother Jones*, January 2013, https://www.motherjones.com/environment/2016/02/lead-exposure-gasoline-crime-increase-children-health/.

282 **Jessica Wolpaw Reyes, wanted to localize:** Jessica W. Reyes, "Environmental Policy as Social Policy? The Impact of Childhood Lead Exposure on Crime," *B.E. Journal of Economic Analysis & Policy* 7, no. 1 (May 2007): 1–43, https://doi.org/10.2202/1935-1682.1796.

283 **lead-crime hypothesis in criminology:** David O. Carpenter and Rick Nevin, "Environmental Causes of Violence," *Physiology & Behavior* 99, no. 2 (2010): 260–68, https://doi.org/10.1016/j.physbeh.2009.09.001.

283 **"The trends have continued":** Rick Nevin, interview with the author, September 2023.

283 **lead dust still embedded in soils:** Howard W. Mielke et al., "Lead in Air, Soil, and Blood: Pb Poisoning in a Changing World," *International Journal of Environmental Research and Public Health* 19, no. 15 (2022): 9500, https://doi.org/10.3390/ijerph19159500.

283 **the United Nations Environment Programme launched:** "Era of Leaded Petrol Over, Eliminating a Major Threat to Human and Planetary Health," UN Environment Programme, August 30, 2021, https://unep.org/news-and-stories/press-release/era-leaded-petrol-over-eliminating-major-threat-human-and-planetary.

283 **countries slowly dropped the fuel:** Hannah Ritchie, "How the World Eliminated Lead from Gasoline," Our World in Data, January 11, 2022, https://ourworldindata.org/leaded-gasoline-phase-out.

284 **Ethyl began diversifying its portfolio:** "The Ethyl Corporation Is Diversifying to Alter Its One-Product Image," *New York Times*, August 15, 1981, 31.

284 **expanded its business overseas tenfold:** William J. Kovarik, "The Ethyl Controversy" (PhD diss., University of Maryland, College Park, 1993), 313.

284 **"I don't recall very much":** Gil Grosvenor, interview with the author, September 2023.

285 **10,000 shares of GM stock worth $2 million:** Email from Collette McDonough, archivist at the Kettering Foundation, to the author, September 22, 2023.

285 **"carry out scientific research":** "Kettering Foundation Names Four Additional Senior Fellows," Kettering Foundation, January 17, 2024, https://www.kettering.org/news/kettering-foundation -names-four-additional-senior-fellows/.

285 **which yielded additional benefits:** Stuart W. Leslie, *Boss Kettering* (New York: Columbia University Press, 1983), 307.

285 **Kettering appeared on the cover of *Time*:** *Time*, January 9, 1933, https://content.time.com/time/covers/0,16641,19330109,00.html.

285 **thirty honorary degrees:** Zay Jeffries, *Charles Franklin Kettering: 1876–1958* (Washington, D.C.: National Academy of Sciences, 1960), 104–22.

286 **a small medical center on the east side of Manhattan:** "History and Milestones," Memorial Sloan Kettering Cancer Center, https:// www.mskcc.org/about/history-milestones.

286 **now considers organic lead:** Hartwig Muhle and Kyle Steenland, "Lead and Lead Compounds," *IARC Monographs*, no. 87 (2006): 12–16.

286 **a carcinogen in mammals:** *Inorganic and Organic Lead Compounds*, IARC Working Group on the Evaluation of Carcinogenic Risks to Humans, vol. 87 (Lyon and Geneva: International Agency for Research on Cancer, 2006).

286 **each generation now appears to have:** "Global Cancer Burden Growing, Amidst Mounting Need for Services," World Health Organization (WHO), February 1, 2024, https://www.who.int /news/item/01-02-2024-global-cancer-burden-growing--amidst -mounting-need-for-services.

287 **many senior officials at the EPA:** Alan Loeb, interview by the author, October 2023.

289 **started offering credentials for people:** "History," Association of Professional Industrial Hygienists, accessed February 12, 2024, https://www.apih.us/.

289 **"I wouldn't change my life":** Barbara Sicherman and Alice Hamilton, *Alice Hamilton: A Life in Letters* (Cambridge, MA: Harvard University Press, 1984), 407.

## Epilogue

292 **Prior court cases have alleged:** The two dominant cases to reference Ethyl's internal discussions about the poisonous nature

of TEL and the profit calculations in its marketing and sale were *United States v. E. I. du Pont de Nemours & Co.* (Case #49-C-1071), first brought in 1949 in U.S. District Court, Northern District of Illinois, Eastern Division. The U.S. Supreme Court (on appeal) sided with du Pont de Nemours in 1956; and *Reginald Smith, Jr. et al. v. Lead Industries Association Inc. et al.* (Case #24-C-99-004490), first brought in 1999 in Baltimore City Circuit Court. The Court of Appeals of Maryland dismissed the charges against Ethyl Corporation in 2002.

**294** **high blood pressure and certain types of cancer:** *Guidance on PFAS Exposure, Testing, and Clinical Follow-Up* (Washington, D.C.: National Academies Press, 2022), https://www.ncbi.nlm.nih.gov /books/NBK584690/.

# Image Credits

# Index

Note: Italicized page numbers indicate material in photographs or illustrations.

# About the Author

**Daniel Stone** is a writer on science, history, and adventure, as well as the author of *Sinkable* and the national bestseller *The Food Explorer*. He teaches environmental science at Johns Hopkins University and is a former senior editor for *National Geographic* and a former White House correspondent for *Newsweek*. His is a Distinguished Smithsonian Fellow at the National Museum of American History. He lives in Atlanta with his wife and two sons.